ESSENTIAL
FIAT 124
SPIDER & COUPÉ

ESSENTIAL
FIAT 124
SPIDER & COUPÉ

THE CARS AND THEIR STORY 1966-85

MARTIN BUCKLEY

Published 1997 by Bay View Books Ltd
The Red House, 25-26 Bridgeland Street,
Bideford, Devon EX39 2PZ, UK

© Copyright 1997 by Bay View Books Ltd

All rights reserved. No part of this publication
may be reproduced or transmitted in any form
or by any means, electronic or mechanical,
including photocopying, recording or in
any information storage or retrieval
system, without the prior written
permission of the publisher.

Edited by Mark Hughes
Typesetting and design by Chris Fayers & Sarah Ward

ISBN 1 870979 99 0
Printed in Spain

CONTENTS

FIATISSIMO!
6

THE SPIDER'S WEB
14

UNITED SPIDERS OF AMERICA
26

ABARTH
39

RALLY TIME
48

THE LOST CLASSIC OF TURIN
61

TODAY
73

APPENDIX
78

FIATISSIMO!

The 124 Spider was the best affordable mass-produced sports car of its generation, the 124 Coupé an elegant sister that was well received in America and throughout Europe – both are seen in early form.

At the beginning of the 1960s *Fabbrica Italiana Automobili Torino*, or FIAT, was known and loved by Italians. The company had produced a long line of competent bread and butter vehicles for its home market, suitably spicing them with a smattering of exotica produced in conjunction with the local design houses and coachbuilders.

Yet outside Italy, and especially in the key markets of Britain and America, the name Fiat provoked only mild interest, especially among enthusiastic drivers. But a decade later, thanks to two cars based on the same ordinary saloon, but with very different development histories, the company had put its name on the map around the world.

It was a triumphal entry by any standards. Both cars, first the 124 Spider, then the 124 Coupé, were right from the very beginning. Not only did they attract massively favourable press comment, but they immediately translated that into huge sales.

The Spider, launched in 1966, was hailed, both then and now, as perhaps the best affordable mass-produced sports car of its generation. It went on to remain in production until 1985, outliving all its contemporary sports car opposition, apart from its rival built by Alfa Romeo.

In 1967, a year later, Fiat followed through with the visually less exciting, but infinitely pretty and competent 124 Coupé, which enabled the company to make a breakthrough in Britain, as well as the rest of Europe. Yet,

Engineering starting point for Fiat's new sports cars was the 124 saloon, seen a few months before March 1966 launch during prototype testing at Vizzola test track and in the Sahara.

despite agonised pleas, no factory right-hand drive Spiders were ever built and the car was never imported into Britain by the factory. Fiat had pitched its sights higher and was looking across the Atlantic to America.

At that time the US sports car market was dominated by traditional British products, with at least three-quarters of the production of MGs, Triumph TRs and Austin-Healeys exported there. Although immensely popular with hard-driving enthusiasts, there were many Americans, used to higher standards of comfort, who yearned for something a shade more civilised – and more modern.

Fiat saw this, and for the lucrative American market produced the sophisticated, yet rugged, Spider with stunning success. Out of 198,000 built, 170,000 were sold there, especially to buyers in rich, sunny states like California and Florida. Meanwhile the pretty Coupé, which also sold in the New World, made an even greater hit in Britain, where initial demand was so high that second-hand examples were selling at a premium over new because of the waiting list.

But unfortunately the execution of both cars never matched up to the wonderful design, which had paid little heed to the perils of damp climates. The Coupé, even more than the Spider, rightly gained the reputation for rust which plagued not just Fiat, but its wholly-owned company of Lancia, and was to give all Italian cars a bad name that took years to shake off.

And for today's enthusiasts that means a sad ending to what is otherwise a happy story. For while Spiders survive in relatively large numbers in America, these are nothing compared to what they would be if the cars had been manufactured from better quality steel and had elementary anti-corrosion measures built in.

Only with later models did Fiat begin to learn its lessons, and for the shorter-lived Coupé, which stopped production in Italy in 1975, that was largely too late. So much so, that of all the great 1960s cars, it probably survives in fewer relative numbers than any other.

Car of the Year origins

The origins of both the Spider and Coupé lay in Fiat's mid-range 124 saloon. This was an important car for the company, replacing the 1100 which had been the backbone of the range since the mid-1950s. Dante Giacosa, the long-serving technical director, had tried to push through a radical, front-wheel drive design, but the car that appeared in April 1966 drove conventionally through the rear wheels. Thanks to careful paring of the unitary structure, this boxy, four-door car was lighter than the Fiat 1100, as well as being faster, roomier and better handling. Just as important from the company point of view, it was designed to be cheaper to build than its predecessor, which, as was normal policy, would remain in production while the new model established itself.

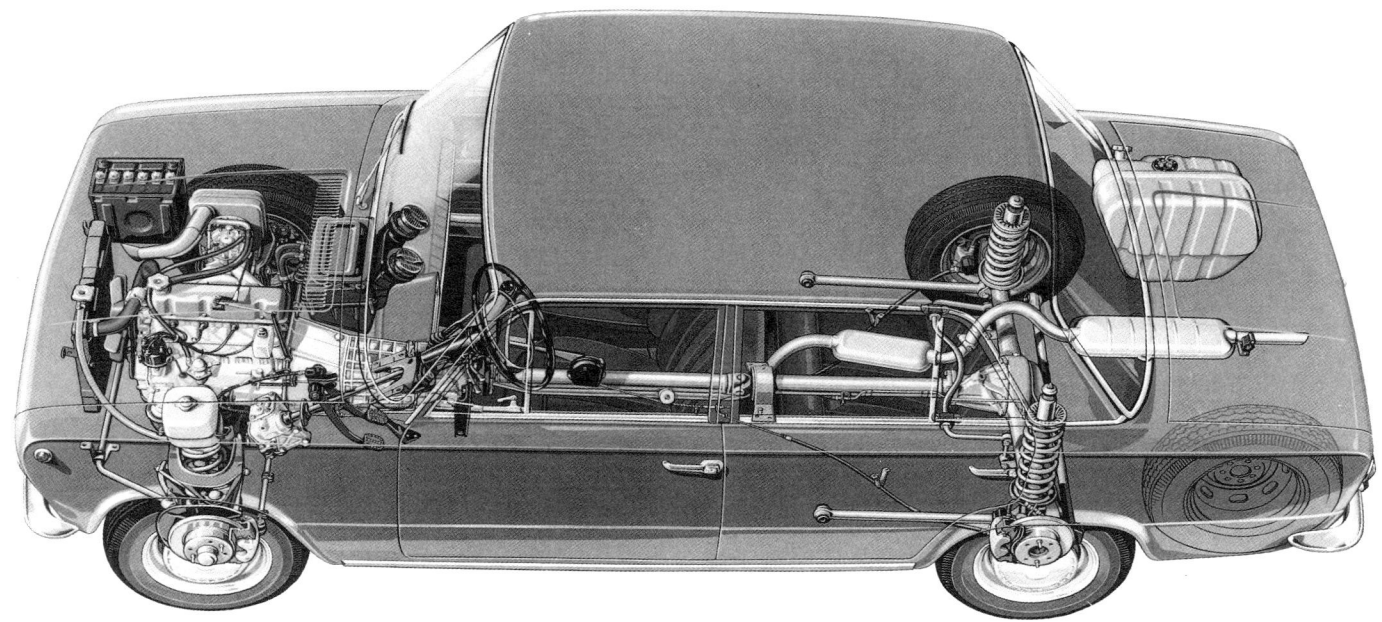

The engine, designed by one-time Ferrari man Aurelio Lampredi, was also entirely new, and the first of Fiat's four-cylinder units to be given a five-bearing crankshaft, with another innovation of a full-flow oil filter, to be changed every 6000 miles. This simple but high-efficiency 1197cc in-line 'four' gave a competitive output of 50bhp per litre, adding up to a rorty 60bhp. Meanwhile its over-square dimensions of 73.0mm × 71.5mm, along with the robust construction of its bottom end, gave obvious scope for enlargement. The alloy head and individual inlet and exhaust ports, with wedge-shaped combustion chambers, were coupled with pistons featuring two pressure and scraper rings and slightly domed to achieve the compression ratio of 8.8:1. Fuel was fed in through a twin-choke Solex or Weber carburettor, mounted on an alloy inlet manifold, while the drive was carried through an all-synchromesh four-speed gearbox with floor change that had been lifted straight from the bigger Fiat 1500.

Weighing in at 1836lb, the 124 could top 80mph and was class-leading on acceleration, achieving 0-60mph in a brisk 13sec, although typically Italian low overall gearing meant fuel consumption was a poor 28mpg.

Fiat eschewed independent rear suspension as too costly and complicated. Leaf springs, however, were too outmoded. The designers therefore located the live rear axle by coil springs and trailing radius arms with a Panhard rod linkage. The entirely conventional front suspension had pressed double wishbones and coil springs enclosing telescopic dampers. The top wishbones and

Although the 124 saloon (above) shared some of its running gear with the twin-cam Spider and Coupé, initially it had to make do with a simple pushrod engine. But in time the sophisticated twin-cam unit found its way into saloons too, first the 125 (below) in 1968 and then the 124 in 1970.

upper ends of the spring/damper units were body-mounted, while the bottom wishbones were linked to the engine-supporting cross-beam, with the arrangement finished off by a stout anti-roll bar. All the bushes were rubber and there were no greasing points.

Although the suspension was conventional, the handling was good. Meanwhile the Bendix type disc brakes on all four wheels, with a single hydraulic cylinder, were an advanced feature for such a modest family saloon. Even better, the brakes were equipped with a pressure valve in the hydraulics to stop the rear end locking up too

Two Fiat sports car ancestors of contrasting character. The 120mph 8V of 1952 was exotic and expensive, the 1100 Trasformabile of 1955 ungainly and unexceptional.

early, a feature first been seen on, of all things, the Lancia Superjolly delivery van.

Nowadays it might be fashionable to smirk at the 124's childish 'three box' profile, but it was roomy, easy to drive, thoroughly affordable and a deserving winner of the 1966 Car of the Year title. 'The most up-to-date solution to the classic layout yet devised,' the authoritative *Autocar* pronounced.

By 1967 Fiat's Mirafiori plant was turning out 1000 a day and had already sold the design, along with the complete factory to build it, to Russia, spawning the ubiquitous Lada, albeit with a different engine. Similar versions for Spain, Bulgaria and Romania were to follow.

The 124 was to remain current until 1974, in several variations. First came an estate, then the 124S, distinguishable by its four headlights and bigger 1438cc engine that increased maximum speed to 96mph. In 1971 the styling was tidied up through new touches to the front and rear ends, along with the addition of an alternator and a servo for the brakes. Best of all for the enthusiast was the introduction of the twin-cam Special T, bringing the sober saloon into the 100mph class, while late – post-January 1973 – Special Ts had an even bigger 1592cc twin-cam and the option of a five-speed gearbox.

Production stood at 1,280,000 in 1974 when the Mirafiori finally took over and Fiat pulled the plug.

It was from this sound base that the Fiat 124 Spider evolved.

Sport of a sort

Although nobody could have guessed the 124 Spider would outlive its sibling saloon by more than a decade, that it turned out to be so good should have come as no surprise. For a giant car maker, Fiat already had a good track record with sports cars. Way back in the first years of the century the company had quickly realised the sales benefits of competition success and forged its reputation with its thunderous chain-driven Grand Prix cars. In the 1930s, for a much wider audience, it had then introduced the slender Ghia-designed 508S Balilla, which not only sold well, but had plenty of competition highlights, including winning its class in the Mille Miglia.

Post-war, as early as 1947 the company was already producing the 1100 Sport, a warmed-up version of the pre-war 508 that soon proved its mettle by taking second, third and fourth in the 1948 Mille Miglia, and also winning its class.

The 8V, introduced in 1952, was on a much more exalted plane – an expensively-built coupé with a high-revving V8 engine and a top speed of 120mph. Rare and exotic, for Fiat's design supremo Dante Giacosa the 'Otto Vu' was a therapeutic diversion from the mundane rigours of austere economy cars and, although few were built, proved Fiat was capable of mixing it with the best. Other big, high-performance Fiats were to follow, most

Fiat in the 1960s developed its intermittent tradition of having a high-performance machine at the top of the range. The elegant 2300S (above) arrived in 1961, the breathtakingly beautiful Dino (left) at the same time as the 124 Spider in 1966.

memorably the 2300S and Dino, along with frugal tiddlers like the 850 Spider and Coupé.

But these were all far removed in spirit from the 124 Spider, a true sports car with much broader appeal, whose true predecessor was the ungainly 1100 Trasformabile. This machine, based on the floorpan of the 1100 saloon and launched in 1955, was actually styled by Fiat, although it bore a certain resemblance to the contemporary Lancia Aurelia Spyder, designed by Pininfarina. Confusingly, Pininfarina built the bodywork for the Trasformabile, which sold well for such a specialised car – 1030 examples of the original 1100cc version, and 2363 of the 1200cc model that followed two years later in 1957.

Immediate predecessor to the 124 Spider was another Pininfarina design, launched in 1959 as the 1200 Cabriolet. Technical development took two directions: the use of an enlarged version of Fiat's pushrod engine (as in the 1500, above) or of a race-bred twin-cam from OSCA (as in the 1600S, right).

Encouraged, Fiat decided to introduce another model based on one of its contemporary saloons, but this time styled as well as built by Pininfarina. The result was the 1200 Cabriolet and Coupé, first previewed in Coupé form at the Turin Motor Show in November 1958, then formally introduced at Geneva in March 1959. Here, with its gently curved windscreen and wafer-thin bumpers, was a svelte and elegant two-seater convertible, featuring the comforts of wind-up windows and saloon car seats. It was coupled with a slim-pillared Coupé version, destined to be built at a slower rate and marketed direct through the Pininfarina sales organisation.

For those who found the 58bhp of the 1200 base model insufficiently poky, there was the option of the 1500, with a new twin-cam engine putting out 80bhp. Unlike the later Lampredi-designed twin-cam, this unit shared nothing with any contemporary Fiat unit, but was instead developed from a racing engine built by OSCA – *Officine Specializzata Costruzione Automobili*. OSCA, founded in 1947 by the Maserati brothers, Ernesto and Ettore, after they had sold their famous racing car concern to the Orsi family ten years earlier, had rarely been financially stable. But the brothers had at last seen a way of securing engines for their own use, without having to invest in expensive mass production tooling, by convincing Fiat to put into production their alloy-

headed, iron-blocked twin-cam. In turn they went on to build a handful of sports cars on the same Fiat chassis, but with more highly tuned engines and elegantly clad in spare, lightweight bodywork by the likes of Zagato, Fissore and Vignale. That, however, is another story.

For its part, Fiat gained a twin-cam design with a racing pedigree that could give up to 110bhp in twin-

Fiat's sports car evolution shot off in all sorts of directions in the mid-1960s. Using the rear-engined mechanical layout of the 850 saloon, this convertible by Bertone – Turin's great rival to Pininfarina – appeared a year before the 124 Spider.

plug, multi-carburettor form. However, to ensure smooth torque delivery, Fiat fitted it with a single twin-choke Weber carburettor that still gave it an impressive 80bhp – around 53bhp per litre.

Outwardly there was little to tell the twin-cam from its pushrod-engined sibling, apart from an off-centre air intake on the bonnet, and there was not that marked a difference in the top speed – 100mph, compared with 90mph. But where the difference really told was in the acceleration figures, with the twin-cam's 0-60mph figure of 15.2sec almost a third faster.

Of course you had to pay for this extra urge – 1,420,000 lira for the 1200, against 1,800,000 for the OSCA twin-cam. In Britain, thanks to the loading imposed by import duty, in 1962 this meant the 1200 sitting in the showroom at £1460, with the twin-cam at £1749. As an MGB sold for £834, and a Jaguar E-type only a shade over £2000, it is not difficult to see why the convertible, the only model marketed in the UK, was such a rare sight.

Production lasted for seven years, with few changes. In March 1963 the base 1200 model became a 1500, gaining the 1481cc engine from the Fiat 1500 saloon, along with front disc brakes and a wider, shallower front grille. From the spring of 1965 there was a five-speed gearbox, later to find a home in the 124 Spider and Coupé. In 1960, a year after production had started, the twin-cam 1500 got front discs and bigger 15in wheels. In October 1962 it was renamed the 1600S when the engine was enlarged to 1568cc and fitted with twin Webers, increasing output by 10bhp to 90bhp to counter stiffening opposition from Alfa Romeo, whose twin-cam Giulia nonetheless always handsomely outsold it. Four-wheel discs and a new front grille arrived in 1963, and from 1965 the same five-speed gearbox as the pushrod 1500. On these later twin-cams the bonnet scoop disappeared, but you can still spot one by its inner set of driving lamps.

Total production figures for the Coupés and Cabriolets had reached 37,000 by the summer of 1966, when production ceased to make way for the 124 Spider. Of this, 11,851 were 1200 Cabriolets and 20,420 1500s, while the total for Coupés was only 2210. Production figures for twin-cams are much less precise, although Fiat historians make an educated guess of around 3000 cars all told. Whatever the true figure, there is no doubt the Cabriolet set a precedent for technical sophistication that was inherited by the 124 Spider when it succeeded it in the summer of 1966.

Two of a kind

Technically, it would probably be true to say the Spider and Coupé had no peers in their class in the mid-1960s. Like their predecessor, their layout was conventional front engine/rear drive, but the details were more advanced. This time there was to be no base pushrod model, but simply one type of twin overhead camshaft four-cylinder across the range.

This engine, schemed by Lampredi, was a classic – sweet, efficient, smooth and punchy, and with a fine, throaty voice that could only be Italian. Yet it was as robust and straightforward to produce as the old OSCA twin-cam had been specialised and finicky. And all at a time when most Fiat contemporaries were equipped with

Slip-road dicing on an *autostrada* in 1974. It is ironic that the Coupé should be the one off the main carriageway, heading for the hard shoulder – it ceased production a year later. That is a twin-cam 124 Special T saloon trying to keep up.

either thrashy, breathless pushrods or big, thirsty six-cylinder units. When this was allied with the early Spiders and Coupés being lighter than both many of their rivals and their predecessors, it added up to topping 100mph with just 1438cc. And there was more. At that time five-speed gearboxes and four-wheel disc brakes were a rarity on anything but the most exalted sporting machinery. Yet the 124 Spider and Coupé gave you both.

Meanwhile the chassis was so well conceived that Fiat proved sports cars did not need to be crude and uncomfortable to handle well. And it was all done without unnecessary complications. The wishbones and coil springs at the front were standard issue and there was nothing more exciting than a live axle at the rear – albeit using coil springs with a Panhard rod, rather than the dated leaf springs of the outgoing 1500/1600S. And even though the steering was not rack and pinion, but a humble worm and roller box like the 124 saloon, it was still praised for its fluidity and precision.

It is amazing to think that critics compared the 124's road manners favourably to that paragon of handling finesse, the Lotus Elan. With hindsight that might be seen to be stretching a point, yet there is no denying the Italian sports car could make a mockery of any Triumph TR or MG on a twisty road, not just because its poise and roadholding was so superior, but because it rode so much better.

It also lasted longer. While the Coupé died out in 1975, the Spider lingered on for a further decade, its twin-cam engine growing to 2 litres and gaining first fuel injection, then turbocharging and, at its last gasp in the 1980s, even supercharging.

Yet it was the styling, not the engine, that enabled the Spider to survive so long. Well proportioned and unpretentious, this was the Pininfarina house at its most chic and accomplished. The Americans just loved the Spider. And they continued to buy it by the thousand, even when it was well past its sell-by date in the late 1970s and 1980s, and had weathered several tweaks, including the indignity of disfiguring impact bumpers and raised ride height. Yet it was never comprehensively restyled in its 19-year production run and commercially remains one of Pininfarina's biggest hits.

Neither was it a car just for hair-shirted masochists. It had a superb water-tight top that could be operated with one hand, a decent heater and even the option of an automatic.

It was fortunate for British Leyland (*née* BMC), who owned the MG marque, that Fiat never felt any need sell the Spider on the British market. The company's argument was that American demand was so strong they could not build them fast enough, while tooling up for a right-hand drive version would not have been cost-effective, especially when there was already no hope of competing on equal price terms with the MGB. In reality the Spider would probably have largely justified any price discrepancy.

The Spider still survives by the thousand in mainland Europe and America, and if you do see one in Britain – there are many in London – it will almost certainly be a personal import, probably from the States. Sadly, though, the Coupé, which once sold in Britain in high numbers, is now almost extinct, so badly and quickly did it rust away.

THE SPIDER'S WEB

An immaculate Spider from 1974, the last year for chrome bumpers. The date makes this a CS type but externally it is indistinguishable from the post-1969 BS, for which the mesh grille was introduced.

There are good years for cars and bad ones. For the Italians in particular, 1966 was a good one. Lamborghini gave us the mould-breaking mid-engined P400 Miura, Maserati the beautiful Ghibli, De Tomaso its brutal but deeply flawed Mangusta, and, unveiled at the Geneva Salon in April, two new Spider protagonists from Fiat.

The Fiat Dino was a low-volume homologation hybrid, fitted with a Ferrari engine, while the 124 Sport – referred to from here on as the Spider to save confusion – was a pretty, but unpretentious, middle-ranking convertible. From being side by side at the show, their ways were soon to part. The expensive, upmarket Dino, boasting a top speed of 130mph, was to cease production in 1973, while the relatively affordable 124 Spider lived on. None of the visitors thronging the stand to gasp at its lines could have ever guessed it would still be on offer nearly two decades later, by which time it would have sold 170,000 examples in North America alone.

And who better for Fiat to choose to style and build it than Pininfarina, by then the biggest and most respected name in Italian coachbuilding? This company, founded in 1930 by Battista Farina – nick-named 'Pinin', an Italian diminutive – had been responsible for some of the most elegant concours d'élégance contenders of the pre-war years, memorably its aerodynamic Alfa Romeos and Lancias and succession of show-stopping dream cars. In the late 1940s and early 1950s its reputation had been further enhanced by classics like the Cisitalia 202, the only car ever to make it into the Museum of Modern Art in New York, and the Lancia Aurelia B20, the first modern GT or *Gran Turismo* car.

Pininfarina also had a long association with Fiat,

This special-bodied Chevrolet Corvette of 1963 marked the beginning of Spider styling themes. The Rondine was designed by American John Tjaarda soon after he left Ghia to join Pininfarina.

starting in 1931, when the coachbuilder had produced a flashy double-phaeton body on the short-lived 2500 model. In 1935 it had begun exploring fashionable aerodynamic themes, with a series of special-bodied 6C 1500 coupés, then, after the war, come up with the neat little 1100 TV coupé.

But although the company had always been based in Turin, it was not until 1958, when it opened its ultra-modern Corso Trapini plant at Grugliasco, to the west of the city, that things really changed. The following year 'old man' Pinin Farina handed over the running of the company to his son Sergio, along with his son-in-law Renzo Carli, and three years later changed his surname, and the name of his company, from Pinin Farina to Pininfarina by Presidential decree. He was to die in April 1966, aged 72, just as the 124 Spider was being announced to the public.

Farina was a personal friend of Ferrari and on the board of his company, and by the beginning of the 1960s most Ferrari road cars were wearing his distinctive badge, as they have done ever since with only a handful of exceptions. Yet at the same time his company had turned its hand just as effortlessly to styling good looking mass-production saloons such as the Peugeot 403 for France and Austin/Morris 1100 for Britain.

Not only Pininfarina was riding the crest of a wave. In the 1950s and 1960s Italian styling had become the height of fashion. The Turin houses of Bertone, Vignale, Ghia and Michelotti had turned themselves into the motoring equivalent of *haute couture* fashion in Paris, with almost all the big European car makers coming along to spice up their latest big saloons or family runabouts with a dash of 'Italian style'.

But it was only Pininfarina, now installed in its new premises, which had the industrial muscle to produce cars in quantity. Usually it made flagship coupés for Lancia and Ferrari, although these attracted few takers outside Italy. But in the 1960s, after styling the regular Peugeot 404 saloon, the company had begun building completed bodyshells for coupé and convertible models, which were then transported overland to France – an arrangement which was later extended to the 504. In 1961 it had even undertaken a short run of special Cadillac Coupés, excruciatingly named the Brougham Jacqueline.

But more than anyone, Pininfarina's main manufacturing customer had become Fiat, whose 1200, 1500 and 1600S convertibles it had been producing in large numbers since 1959 and for whom it now began work on the design of the 124 Spider.

Although Farina's son, Sergio, had been responsible for design at the company since the 1950s, a young American, Tom Tjaarda, was given the job – understandably in view of the market the car was to be aimed at. Tjaarda was the son of the maverick Detroit stylist of the 1930s, John Tjaarda, and had been studying architecture as a graduate of the University of Michigan when he had been head-hunted by Ghia and brought to Turin in 1958. The following year, at the age of only 25, he had styled the Fiat 2300S Coupé and Innocenti Spyder 950, before resigning after an argument and joining Pininfarina in 1961. He stayed there until 1967 and is credited with the Ferrari 330GT and California Spyder, as

With Rondine-like reverse-rake rear pillars evident, this is Pininfarina's 1966 proposal for a 124 Coupé – essentially a Spider with a fixed roof. But at this stage Fiat was finalising its own Coupé design.

A Spider prototype testing on an *autostrada* near Turin in April 1966. Disguise panels conceal the shape of the nose, make-do tail lamps are fitted, and the soft-top lacks the quarter windows found on production cars.

well as elegant one-offs like the Fiat 2300S Lausanne and a series of special-bodied Lancias and Chevrolet Corvair coupés. After that he returned to Ghia, where he is now best remembered as the man who styled the De Tomaso Pantera – and the original Ford Fiesta.

One of his first jobs for Pininfarina had been a special-bodied Corvette called Rondine, a one-off show car first seen at the Paris Salon in 1963. When, in the same year, he began his 1/5 scale models for what was to become the 124 Spider, it was the Rondine, more than anything, that influenced the shape first spotted by the Italian motoring press three years later when it was being tested on an *autostrada* near Turin.

Nearly identical to the Rondine was the character line that swept from the middle of the front wing and then kicked up above the door handle to form the top of the rear wing. From the rear, very similar, although less exaggerated, was the inward slanting boot and rear panel, while there were strong Rondine echoes in the roof line of the optional hardtop, although the reverse-rake rear window, in the style of the Ford Anglia, had been abandoned.

There were also touches of contemporary rivals. The slim grille with triangulated ends hinted at the Datsun Fairlady, the recessed front lights at the MGB. Pininfarina as a house owned up to the latter – it had tidied up the shape of the B for 1962. The resemblance to the Fairlady was also not so surprising. Several Japanese cars of the 1960s were actually styled in Italy, but sometimes the manufacturers kept this quiet in order to save face – and the Fairlady may well have received some covert Pininfarina input.

At the front the driving lights and indicators were tucked snugly in below the main headlamps, with rear brake and turn lights enclosed in a slim moulding at the rear. The front-hinged bonnet swept smoothly down between the crown of the wings, its leading edge forming the top of the grille. The only Fiat badges were a laurel wreath, repeated on the boot lid, along with a discreet 'Fiat 124' script on the rear panel. Meanwhile Pininfarina crests were carried on the rear wings, between the door and wheel arch, to signify that the Spider was not just styled by Pininfarina but built, painted and trimmed in its factory on a drive train assembled by Fiat.

Initially Pininfarina planned to build up to 60 examples a day, little knowing that by the late 1970s, at the height of the Spider's popularity, they would be producing 150 a day.

In profile the 124 Spider was elegant and well-balanced, looking soft and inoffensive in contrast to the swooping, muscle-bound lines of the Dino. This, according to the magazine *Style Auto*, was quite intentional: 'The Dino has been designed mainly for men: the 124 is intended for a very heterogeneous clientele, for the middle classes for whom it would have the role of a second car, probably for the wife. It is not difficult to imagine what would happen to a model which, because of its too masculine appearance, was not popular with women.'

To nullify any implication that this feminine softness meant weakness, Fiat claimed the structure it supplied for Pininfarina to build on was more rigid than that of the saloon. The Spider was certainly heavier, all told, at 2090lb, compared to 1930lb. But at the same time its body was more luxuriously equipped, while the nature of its construction, featuring hand finishing and therefore the use of lead to form smooth shapes around arches and joins, meant it was bound to be more portly.

Although 5.4in had been clipped from the saloon's wheelbase, the Spider, with more overhang at the rear, was both longer and wider. Although the open car had a larger 11.8 gallon fuel tank, weight distribution – a slightly nose-heavy 55%/45% – was identical. The car rode on innocent looking 13in steel wheels with plain hub caps, shod with modest Pirelli Cinturato or Michelin X 165 radial tyres.

Brightwork was sparingly applied – a slender one-

More testing: a pair of pre-production cars at Fiat's proving ground.

piece bumper (without over-riders) at the front and sketchy quarter bumpers at the rear, split by the number plate holder.

The soft-top, folded down, hardly intruded into the profile and was a minor work of art in operation. It could be lowered quickly and easily from the driver's seat, using just one hand, and featured glass rear quarter lights to provide good visibility. Over-centre clips, similar to those on the Alfa Romeo Spider, were so easy to release that you could almost raise or lower the top in the time it took for the lights to go green. It also looked good when it was up, which was something that could not be said about many of its contemporaries.

Power with pedigree

Under the bonnet the 124 Spider was no less impressive. Fiat made no attempt to get away with a tweaked version of the pushrod saloon engine, but instead delighted visitors to the 1966 Turin Show with a handsome new 1438cc twin-cam. The design was by Aurelio Lampredi, best remembered for his work at Ferrari on 375 series V12 engines and the World Championship winning 2-litre units of 1952 and '53. He had joined Fiat in 1955 and quickly gone to work raising the power output and efficiency of the production engines, as well as developing the OSCA twin-cam fitted to the previous 1500S/1600S models. Although this engine certainly influenced him, the new twin-cam shared nothing with it.

Lampredi had also been responsible for the standard pushrod unit fitted to the 124, but unlike Lotus, which had used a production pushrod engine (from Ford) as the basis for its own twin-cam, his unit was virtually all-new.

Spider in its earliest AS form, with slatted grille and slender tail lights. The car has a far better contrived soft-top than most British rivals: it can be lowered quickly from the driver's seat, using just one hand. Shiny black plastic cabin, so typical of 1960s sports cars, has a cheap look today, but wood for the dashboard and steering wheel tries to raise the tone. For a driver's car, though, the important thing is a clear and comprehensive set of dials.

The only major component in common with the pushrod engine was the beefy, fully balanced, five-bearing crankshaft, although bore centres were shared as the twin-cam's iron block had to be machined on the same tools as the saloon's pushrod.

To British eyes the twin-cam layout, with all its advantages in terms of gas flow and free-breathing efficiency, might have seemed extravagant. But not to Italians. Alfa Romeo not only fitted its sweet twin-cam 'four' into all manner of boxy saloons that competed dangerously closely with Fiat's mainstream offerings, but even installed it in a van. Fiat, in turn, by utilising existing

Designed by ex-Ferrari man Aurelio Lampredi, the Spider's new twin-cam engine was a jewel. A significant novelty was the use of a toothed rubber belt to drive the camshafts. Power for the first 1438cc version was quoted as 90bhp.

124 engine tooling, made the twin-cam almost as cheap to produce as the pushrod, enabling it to find so many broader commercial applications that it has survived well into the 1990s, used on Lancias.

The cylinder head, with its pent-roofed combustion chambers and inclined valves, was an aluminium gravity die-casting with the valve seats inserted. The plugs sat in line in the valley created by the cam boxes, set in the rear side of the chambers on the engine centre line. Valve timing was different to the saloon, with 52 degrees of overlap compared to a milder 42.

Meanwhile the cam boxes acted as carriers for the bucket tappets, which featured a new time-saving (and thus money-saving) method of valve clearance adjustment. In view of the large numbers Fiat planned to produce, shim adjustment would have been too laborious. Previously, on the engine production line, the clearance between the tappet faces and the cam were measured, the camshaft removed and shims of the correct thickness then inserted between the valve stems and the buckets. But on this engine the cam boxes, complete with camshaft tappets, were supplied as an assembly and attached straight to the head, which was already assembled with valves and springs. Instead of being inserted beneath the tappets, the thick shims sat in recesses on the top faces and were then inserted by depressing the tappet with a special tool that slipped them in with magnetic forceps. This explained the angled faces of the cam covers, which made it easier for an owner or mechanic to insert the levers and shims under the camshaft, saving a great deal of time and meaning a great breakthrough on a mass-produced engine, as valve clearances could now be adjusted without disturbing the timing.

Following the lead taken by Glas of West Germany, Fiat then used quieter, cogged camshaft belts manufactured from rubber reinforced by glass-fibre. These were cheap to produce and, 30 years on, are virtually universal on mass-produced cars, but they were something of a novelty in 1966. Drive came straight off the crankshaft nose, round a smooth jockey pulley tensioner that bore on the back of the belt and pushed it into contact with the jackshaft pulley, which provided the drive for the distributor, gear type oil pump and diaphragm petrol pump. The belt then passed over the inlet and exhaust camshaft pulleys and back to the camshafts over the jackshaft pulley. Fiat recommended the belt be changed every 24,000 miles (early engines without an automatic belt tensioner) or 36,000 miles (all later engines), and prudently provided a plastic cover to protect prying fingers or loose neck ties.

The twin-cam had a slightly bigger sump than the pushrod unit and featured a centrifugal oil filter with a bypass connection. The electric thermostatic cooling fan was a rare fitment in the mid-1960s, while the 12-volt electrics were powered by an alternator, rather than the dynamo which was then still the norm on mass production cars.

Carburation was by a single Weber 34 DFH or DHS/1 with a progressive secondary choke, mounted on a water-heated plenum chamber in the manifold, with four separate curved ducts to handle the mixture. The double exhaust manifold was an elegant casting which, to prevent cross-feeding, took waste gases separately from the outer number 1 and 4 cylinders and the inner 2 and 3 pairing. Power output was high for a 1438cc 'four' at 90bhp – up 31bhp on the saloon and 10bhp on the 1600 OSCA twin-cam. But torque was on the weak side, only 79.6lb ft peaking at 4000rpm.

This did provide a good excuse, however, to make full use of the five-speed, all-synchromesh gearbox, which almost matched the Alfa Romeo 'box for slickness, enabling the driver to delight in keeping the twin-cam in its optimum range. Lest we forget, at that time five speeds were a rare and exotic refinement, usually reserved for high-priced machinery from the likes of Ferrari, Aston Martin or Lamborghini. The 'box was taken from the saloon, with a special tail section added, and the ratios were longer, with fifth very much an overdrive and maximum speed achieved in fourth – although at 18.3mph per 1000rpm this was hardly inter-galactic. Drive was taken up smoothly by a 7.9in diaphragm sprung clutch.

The suspension followed saloon practice, with wishbones and coil springs at the front, plus telescopic shock absorbers. To provide more roll-free cornering, spring rates were raised by 19% at the front and 14% at the rear, and the front anti-roll bar thickened from 20 to 21mm.

At the back was the same lightweight hypoid bevel live rear axle as in the saloon, located by similar parallel trailing arms, coil springs and Panhard rod. There was another anti-roll bar and a torque tube on the nose of the axle to resist wind-up. Meanwhile the saloon's 8.9in solid disc brakes with a pressure-limiting system on the rear wheels remained, but beefed by a Master-Vac servo.

At the wheel today

These original Spiders are still impressive when you climb aboard one. The interior decor, with all its shiny plastic, might look a little uninviting 30 years on, but the layout of the dash is still excellent. The easily read Veglia circular white-on-black instruments include a 120mph speedometer, a tachometer reading to 8000rpm, and smaller fuel level, oil pressure and water temperature gauges, all set in a slightly penny-pinching real wooden facia. Other functions are mounted on column switches or scattered around the centre console.

The plastic seats are contoured well enough and in addition you can adjust the backrest, a rare feature on sports cars in the 1960s, while the driving position is not as Italian as many of its compatriots, even if the straight-arm stance, wheel well forward and slightly horizontal, is fairly typical. There is plenty of leg and elbow room and, unlike so many competitors, the footwells are wide, reflecting the fact that the Spider was a good 5in wider than most of its sports car contemporaries. The pedals, set close together, are not too far forward, while the stout and gaited gear lever emerges from the transmission tunnel at

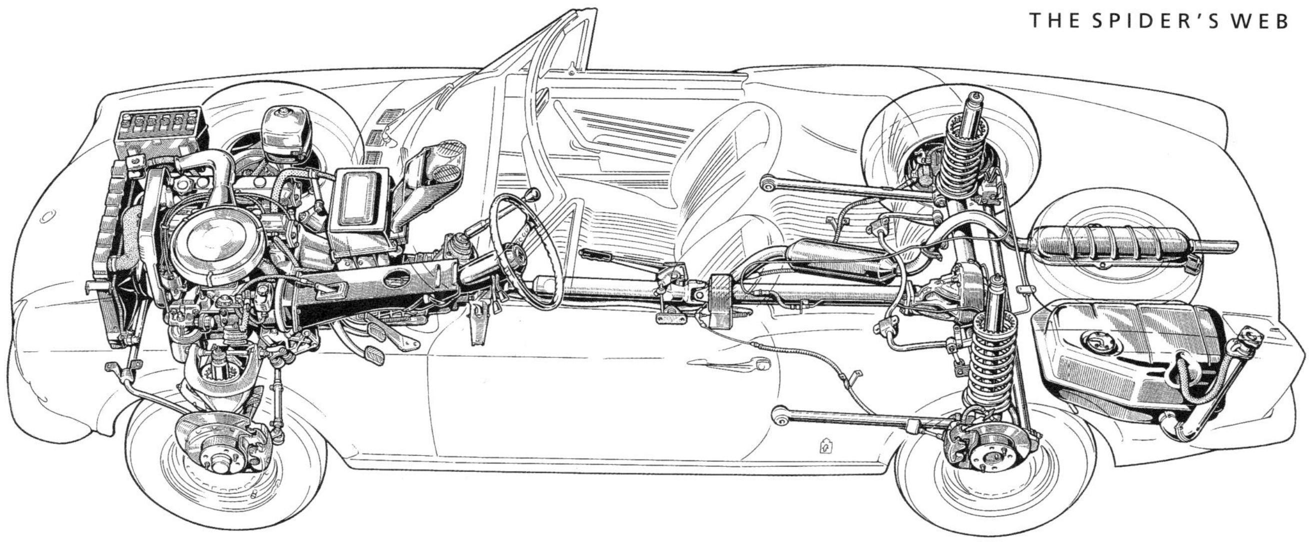

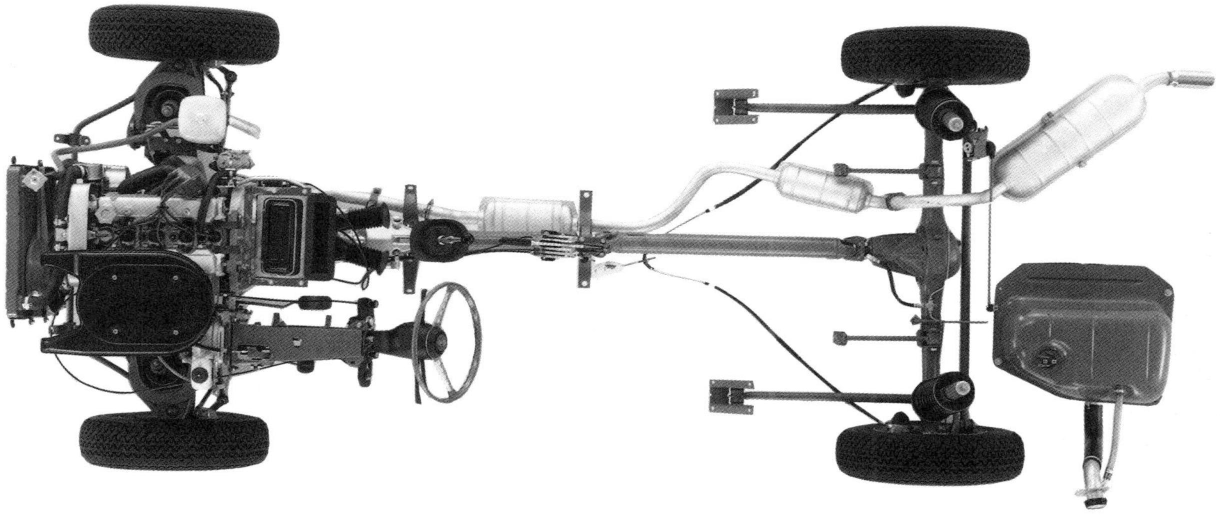

THE SPIDER'S WEB

Much of the Spider's underpinning came from the 124 saloon, with use of a live rear axle the only unsophisticated aspect. Double wishbones (with anti-roll bar) form the front suspension and braking is by discs all round.

a conveniently slight angle. The steering wheel is thin-rimmed, with drilled spokes.

The soft-top, the best in the business, gives more light and better visibility than almost any contemporary, thanks to its integral rear three-quarter windows, and drops neatly down inside the rear wings when folded.

If the engine is in good shape it starts easily, hot or cold, with either a couple of pumps on the throttle, or a little choke, and warms up quickly. It always has a fairly lumpy idle, but sounds eager – fruity yet refined – with no cam drive noise and only a hint of tappet click, thanks to those toothed belts. The clutch is smooth and not

ESSENTIAL FIAT 124 SPIDER & COUPÉ

Comparative views show Spider's three configurations: soft-top stows neatly to sit flush with the body; quarter windows give good visibility with soft-top raised; and hard-top's wrap-round window also maximises driver's rearward view.

US advert has simple copy and a striking image: the Spider almost sold itself when it went on sale in America in 1968, two years after launch in Europe.

Match this. $3181: **FIAT** 124 Spider

Brochure images from 1972, when the Spider received the revised 1592cc and 1756cc single-carburettor engines introduced with the 132 saloon. Simple styling tweaks – a new grille, twin bonnet bulges and larger tail lights – had arrived for the BS model in 1969, and striking Cromodora alloy wheels were an option.

Fiat 124 Sport spider 1600/1800

	"1600"	"1800"
Motore:	1592 cm³	1756 cm³
Velocità:	180 km/h	185 km/h

FIAT

Small revisions were made to the interior in 1969: an extra instrument on the dashboard, matt black finish for the instrument bezels and steering wheel spokes, and a redesigned centre console with wood trim and a more centrally positioned gear lever. The head restraints were optional.

excessively heavy, and the gear lever slices quickly through the ratios, with a well-defined gate and spring loading towards third and fourth. A diagonal push gets you into fifth, in the top right-hand corner, and while the synchromesh is unbeatable on the way up, you can make the rings baulk a little if you come back down too vigorously.

Pulling power is so mild at low revs it is better to use the gears and a generous amount of the long, smooth throttle movement to take you up to the red line, set at 6500rpm. Give the twin-cam its head, though, and it will go on to 7000rpm, and even beyond, without sounding harsh or strained.

First takes you to 30mph, second to just under 50 and third to almost 75, while the maximum of 104mph is achieved in fourth, with the overdrive fifth good for cruising at up to 100mph. In reality 70-80mph is a more realistic pace, with just a quiet hum from the engine and exhaust. At speed, with the top down, there is a little buffeting, but you can still enjoy the elements without having to wrap up like an Eskimo, even in chilly but bright weather, thanks to the efficient heater with its three-speed fan.

In its prime, the original 1438cc Spider would sprint to 60mph in 11sec and touch 30mpg if driven gingerly – figures that MGB owners, packing 1800cc and a lot more torque, could hardly match, let alone better.

If the sparkling performance of the engine is the Spider's most impressive aspect, its chassis comes a close second. The steering, although a mere worm and peg box, is light, yet direct and reasonably high geared, feeding back everything you want to know about what is happening to the modest 5in wide tyres. Understeer is tyre-scrubbingly strong, but body roll moderate, and if the grip is not ultra-tenacious, the poise and predictability of the chassis make the car a delight to drive on a twisty road.

Bumpy surfaces never worry the car. It has a stiff, rattle-free structure, coupled to springs with a generously long travel rate that mean axle hop on rough corners is minimal and the ride remarkably smooth over ridges and potholes, thanks to damping that is firm on the rebound. The brakes are strong, yet light to use and resilient enough to pull the car up in a straight line time after time.

This, then, is the original 'AS' series Spider, recognisable by its slatted grille. A little under-powered perhaps, yet more than making up in spirit and personality what it lacks in ultimate urge. Unlike so many of its ageing rivals, the original 124 Spider was a true sports car for the 1960s, rather than a warmed-up design from the 1950s.

AS to BS

Fiat could not make enough 124 Spiders, particularly after the car went on sale in America in 1968 and production had to double. Perhaps the only affordable sports car rivalling it was the new Alfa Spider, built alongside on the Pininfarina production lines. This matched the Fiat through the gears, but had a higher top speed. Yet its handling was more involving, its roadholding perhaps a little less good and its ride certainly more busy. It was also less conventionally handsome, although it was to prove more enduring, outliving the Fiat by more than five years.

There were no worthy British contenders, even if the Triumph TR5 (TR250 in the US) had the attraction of a meaty straight-six engine, while Porsche had yet to

Two publicity shots put the Spider into oddly cold context. The couple in shades present some sports car romance, but does the 'Ice Maiden' image make you want this car?

announce its mid-engined 914. Yet even this German rival never seriously challenged the Fiat, despite the cornering power bestowed by its mid-engined layout and the attraction of Teutonic fit and finish. In contrast to the Fiat, it was slow, decidedly expensive and, to many, lacked the aesthetic appeal of a traditional open-topped roadster with a long bonnet and short rear deck.

There was only one major change to the Spider AS series during its run. In September 1968 Fiat quietly ditched the torque tube across the whole 124 range and substituted two extra, short radius rods, fitted above the axle tube towards the centre of the differential. Breakages were coming to light on rough roads with the original system, as the torque tube would exert substantial loads on the axle casing unless the radius rods were exactly aligned.

The 124 BS model arrived in late 1969. It could be spotted at a glance by its honeycomb grille, while other detail changes included new positioning for the laurel wreath badge in the centre of the grille, chrome strips on the sills, and larger rear lights. Inside it had matt-black finish instruments and steering wheel spokes, plus a new centre console and heater controls.

Although a 1438cc model was still offered, the big news was the availability of the 1608cc twin-carburettor engine, as used in the 125 saloon. The new capacity was achieved by increasing the stroke to 80mm, giving the engine exactly square cylinder dimensions. At the same time it differed from the 125 in using twin Weber carburettors, raising power to 110bhp at 6400rpm on the European version, which also had twin bulges in the bonnet, as if to signify the presence of the bigger engine.

UNITED SPIDERS OF AMERICA

Big bumpers are the giveaway to this Spider's identity as a post-1974 model designed to satisfy American regulations – in fact it is a 1984 Pininfarina Spidereuropa, from the period when the Turin design house had taken over marketing of the car from Fiat.

What originally inspired Fiat to build the 124 Spider was the sheer size of the American market. The special needs of this massive country shaped the car's development during the whole of its 19-year production run.

A mere 21 Spiders had been built in 1966, the first year of production, rising to 5500 in 1967. But from 1968, the first year they were imported into the USA, production soared, reaching over 15,000 in 1974 and then, after a slight trough, peaking at 19,931 in 1979. Out of the 198,000 produced in all, a staggering 170,000 were sold in America. But although these numbers may look huge, it should still be remembered that, by Fiat standards, the Spider was low-volume fare, generally produced at the rate of just 30 to 60 a day.

There had been great enthusiasm for European sports cars in the States since the late 1940s, when the first left-hand drive MGs had been imported. With no native competition, it was a market ripe for exploitation that no sports car manufacturer could afford to ignore. Enthusiast magazines like *Road & Track* egged on the demand for these hard-riding, rorty Europeans, with their sidescreens and nimble handling, and while marques like Mercedes, Jaguar and Porsche fought it out at the expensive end of the market, British Austin-Healeys, Triumphs and MGs competed for mid-range volume sales, particularly in hot, sunny markets like California.

Lancia and Alfa Romeo had already built open-topped cars especially for the American market. Although Fiat had been keen to join in, it had had no real contender until the 1200 and 1500 models preceding the 124, which had sold a respectable 37,000 world-wide, although it had established a foothold in the lower echelons of the market with the 850 Spider, a rival to the Triumph Spitfire and Austin-Healey Sprite.

When the 124 Coupé and Spider finally arrived they were given an enthusiastic reception, especially on price. The Spider, complete with 12-month or 12,000-mile warranty, retailed at $3265. This meant it undercut the Triumph TR250 (the TR5 stripped of its fuel injection) by $200 and was only marginally more expensive than the MGB and Datsun 2000 Fairlady. Furthermore it was offered in a range of seven colours marketed under delightfully exotic Italian names – Rosso Corsa (bright red), Bianco (white), Verde Scuro (green), Azzuro Pervinca

UNITED SPIDERS OF AMERICA

America was the only destination for the Spider between 1975-81, but thereafter sales resumed in mainland Europe too. Publicity images from the US and Germany both date from 1981 – but wheel styles differ.

(light blue), Blu Medio (medium blue), Nero (black) and Grigio Medio (yellow).

On top of all this, it was seen as more grown-up and substantially more refined than its mostly British opposition, while its natural rival, the Alfa Romeo Spider, cost almost $1000 more. On the other hand, a mid-range, full-size American saloon such as the Chevrolet Impala could be had for the same price, making the Fiat a car for the comfortably off. Many American Spiders were bought as second cars, or as graduation presents by rich parents.

A *Road & Track* owner survey, published in 1970, confirmed that the original 1438cc Spider was bought mainly for its looks and handling, but then criticised for its lack of torque. The 1608cc version was therefore a welcome improvement when it arrived in 1971, except that in reality Fiat was just running to stand still. American buyers had to make do with just 104bhp on their de-smogged, single Weber equipped imports. These may have been given a higher top speed, and more flexibility, but they still showed no improvement on acceleration, despite closer gear ratios. And with a price tag of $3572, they now fell between the elderly MGB and the new Porsche 914. Fuel consumption had decreased to around 23mpg and critics complained that the handling had lost some of its sharpness, with more understeer and body roll, and brakes that were now less fade-resistant, as well as over-servoed. The 1968 revisions to the rear suspension were probably to blame for the loss of handling finesse, but it would be wrong to overstate the case. By most standards the Spider still handled superbly.

Meanwhile in Britain the 124 Spider remained basically unknown. Even in the face of clamouring demand, Fiat still steadfastly refused to build the car with right-hand drive, despite the bodyshell carrying blanked-off holes for right-hand steering column, idler box, brake servo and pedal box. Only the enterprising London dealers, Radbourne Racing, followed by Huxford in Hampshire, converted a handful of used Spiders imported from Europe, utilising steering boxes and idlers which they took from the coupé. Apart from that there was only Burlington's of Camden in North London, who had a small sideline in rebuilding damaged Spiders imported by US servicemen serving at English air bases, and converting some of these cars to right-hand drive.

Back in Italy, meanwhile, changes continued to keep the Spider in line with contemporary Fiat saloons. In August 1972 two new engine capacities were offered, both based around the castings found in Fiat's new middleweight saloon, the 132. A new short-stroke version gave 1592cc, which usefully dropped the capacity below the Italian tax break of 1600cc, while a more frugal and easier-to-tune single Weber carburettor returned. Yet, at 108bhp, the 1592cc engine was 2bhp down on the 1608cc, and in addition lost much of its free-revving sweetness thanks to less sporting camshafts. Few saw the point and it was dropped only a year later.

The other new engine, a 1756cc unit designated CS1, was so strangled by emission controls that it only bettered the old 1608cc by 8bhp in European form, while in the USA, due to the catalytic converter fitted to the exhaust, power was down to 87bhp. The five-speed gearbox was a revised unit with closer ratios and a remote linkage (giving a shorter lever), while the optional four-speed was created by simply chopping off top gear on the old type five-speed. The final drive was also lower, at 4.3:1.

Yet, despite the power loss, sales were as strong as ever – so good that in 1975 Fiat began producing the car exclusively for the American market. There would be no more European Spiders until 1981, although it was possible to have a car built to special order and then collected straight from the factory.

In order to comply with Federal impact regulations, US specification cars were then equipped with jacked-up ride height and ugly bumpers attached to hydraulic rams, although to be fair to Fiat they solved this aesthetic conundrum better than most, facing the chrome tubes with hard rubber. But these devices still added 350lb to the weight of the car, which, combined with the low-powered engines, made for gutless performance – Federal Spiders could only struggle to 95mph and had acceleration inferior to the original 1438cc model.

As well as spoiling the looks by putting too much space between tyre and wheel arch, the change in ride height increased roll. Yet the American magazines still reported favourably on the handling, as well as the curvaceous styling that still seemed so tirelessly fresh and pretty, especially when contrasted with Triumph's wedge-shaped TR7 or the disfigured rubber-bumpered MGB. As well as the outsize bumpers, the Federal Spider can be spotted by its rectangular side marker lamps.

Under the bonnet everything looked much as before, except that, inspired by the Abarth rally cars, the battery had moved to the boot.

One advantage for the Spider as an affordable drophead with sporting pretensions was that its rivals were already fading away. The MGs and TRs were on their last legs, there were no more open Datsuns, and the Alfa was still viewed as a more expensive and up-market production. In 1968, when the first Spider went on sale, American buyers had had a choice of 31 different soft-top cars from overseas manufacturers. By the beginning of the 1980s that had shrunk to just five.

Spider fans were given a little to cheer about in July 1978 when a stroke increase to 90mm brought capacity up to 1995cc. But although the Spider 2000 CS2 had more torque – 104lb ft at 3000rpm – it only put out an extra 3bhp. And in very heavily de-smogged Californian form, power was actually down again, to a weeny 80bhp. Yet a combination of a lower final drive ratio and the extra torque improved acceleration across the board, even if the engine was less sweet, especially above 4500rpm. To comply with safety regulations the speedometer was doctored so the needle stuck at 85mph, but contemporary tests showed the car was good for 102mph.

Suspension settings were much as before, but the back axle was now a hybrid of a 124 casing, 131 saloon differential and halfshafts with 132 caliper mountings.

Meanwhile, there was a new option of a Strasbourg-

UNITED SPIDERS OF AMERICA

A regressive phase for the Spider aesthetically and dynamically: the 1975 model, produced exclusively for the US, required side marker lights, '5mph' bumpers and raised ride height to place those bumpers at the prescribed level.

Propelled by American demand, Spider production reached this total during 1975. Celebrations are taking place at the end of the Pininfarina assembly line, but the lack of wheels confirms the convoluted manufacturing process – finished bodyshells were then sent to Fiat for mechanical parts to be fitted.

sourced GM three-speed automatic transmission, as found in the Opel Ascona 1700, lashed up to a lower axle ratio and a separate transmission oil cooler. This knocked a couple of mph off the top speed and added a couple of seconds to the 0-60mph figure, yet did not affect mpg. Outwardly there were bigger bonnet bulges and a spoiler underneath the front bumper, while the tail lights had grown and the doors now featured elegant lozenge-shaped inset handles, as on the Ferrari 400, whose body was built in the same Pininfarina factory.

The restyled steel wheels, shared with the new 2-litre 132 saloon, still wore slender 165 tyres. But buyers could now opt for Cromodora alloy wheels with wider Pirelli P6 tyres. This lower profile rubber made for heavier steering, but ensured the handling was even more forgiving, if not so entertaining. They also gave enormous grip, although eventually an inside tyre would lift and spin as it ran out of traction. A taller soft-top, still as convenient to use as ever, increased headroom but also meant more wind noise. And for the first time Fiat began addressing rust problems by fitting plastic liners to the front arches.

In May 1980, in a search for improved economy, better cold starting and cleaner emissions, Fiat introduced Bosch L-Jetronic electronic fuel injection, lifting power to 102bhp on an 8.2:1 compression ratio and improving fuel consumption to 25mpg, or as much as 35mpg if driven carefully. Top gear was now lower, but timid valve overlap made the twin-cam so breathless beyond 5200rpm that healthy torque and the close intermediate gear ratios were now the secret of the still respectable performance. *Road & Track*, after recording a 0-60mph time of 10.9sec,

An American Spider from 1977, shortly before the 1800 engine was superseded by a 2000. Interior continues to show little change from the design initiated 11 years earlier, although tan trim is more appealing than the universal black of the early days.

UNITED SPIDERS OF AMERICA

Whereas headlamps and grille changed at the front, rear aspect of 1975-81 model differs only in those obtrusive bumpers. Details show the designer's badge and the two styles of exterior door handle, the change to the flush design having occurred in 1981.

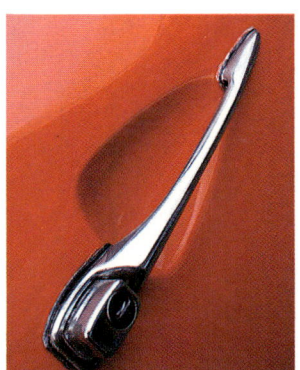

coaxed the now portly 2380lb Spider up to 109mph, making it rather quicker than the 2-litre Alfa Romeo.

Inside, the seats had more lateral and lumbar support and adjustable head rests, with the option of leather, along with electric windows. The wood-rimmed steering wheel disappeared, replaced by chunky leather, although oddly the gear lever now had a wooden knob. There was a new 'fasten belts' warning chime when the ignition was switched on, while the belts themselves were colour co-ordinated with the seat trim.

The next year, 1981, saw a new development when Fiat Motors of North America sanctioned the production of an after-market turbocharged model, to be fitted at the port of entry in the US by Legend Industries, with the Fiat warranty still applying and all parts available from local dealers.

The Warner IHI RHB6 turbo, made by Ishikawajima-Harima Heavy Industries, weighed 13lb (without wastegate) and, with its small-diameter turbines, promised minimal lag. Legend Industries, who spent 18 months developing and testing the car, were looking to give the Spider the feel of a big-engined car, rather than its normal

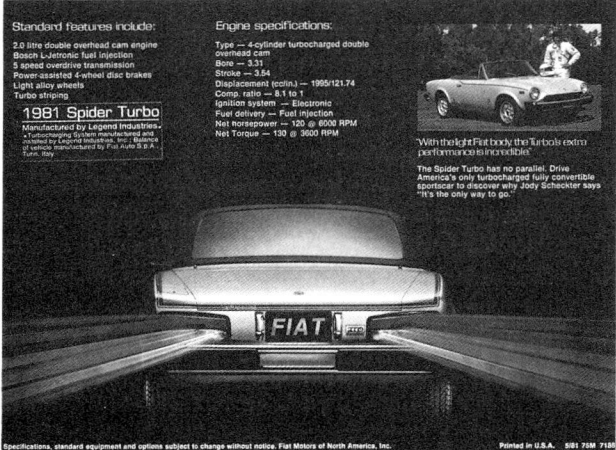

A wild development out of America, not Italy. In 1981 Legend Industries offered an after-market turbo version of the normal injected 2000 Spider, with power uprated from 102bhp to 122bhp. Around 700 were sold.

frantic top-end thrust. The company also hoped to minimise strain on the otherwise unchanged bottom end, so although peak boost, wastegated to 6psi, came in at just 3000rpm, the benefits could be felt from as low as 1400rpm. The turbo stayed working with 60% efficiency at 6000rpm, so there was no longer much point in reaching for the red line.

The turbo lifted the 102bhp of the standard engine to 122bhp, but improved torque to such an extent that Legend estimated it to produce more torque between 1900-5500rpm than the standard car did at its peak. The turbocharger bolted straight onto a special exhaust manifold, nuzzled tight against the right of the block and close to the exhaust manifold, so the gases had a short journey. The wastegate valve and turbine bypass were built into the turbine housing, and activated by a rod linkage connected to a boost-sensitive diaphragm capsule, bolted to the side of the compressor housing.

Because the boost pressure was modest, Legend was so confident about reliability that the basic engine sported few of the modifications found on contemporary turbo installations, doing without forged pistons, sodium-cooled valves, oil cooling or uprated valve guides, while the cooling system was stock and even the compression ratio the same. Fuel injection was retained, although to sustain boost the system was tweaked so the mixture would be enriched as the turbo began to spin hard and back off the timing.

The upshot was smooth, instantly accessible power in each of the five nicely spaced gears, which ran you up to 106mph in fifth. Power came on from as little as 1500rpm and the engine was so flexible third gear would take you from 17 to 75mph. The 0-60mph time of 9.5sec (8.6sec

Having built bodies for the Spider since the beginning, Pininfarina also took over full assembly and sales in 1981, the car ceasing – after a transition period – to bear Fiat identity; the company's wind tunnel is seen in the background. Refreshed interior design, here with optional automatic transmission, came in 1982.

according to *Car and Driver*, which reported quicker times for the prototype automatic model than the manual) was a worthwhile improvement over the standard injection model but still did not tell the whole story, as the blown car had a much more authoritative kick beyond 80mph.

At $14,995 (without optional air conditioning, leather upholstery or even a decent sound system), the Turbo was $2000 more expensive than the stock injection model. But you did get coachlines and Turbo badges, along with the 5.5J × 14 Cromodora alloys, shod with Pirelli P6s, which were later to become standard on the ordinary Spider. Yet despite enthusiastic testimonials from the motoring press, as well as Grand Prix driver Jody Scheckter, who featured in advertisements, it was not a success and in the end only 700 were sold.

For what was now a 17-year-old design, meanwhile, the Spider was wearing surprisingly well. To celebrate its enduring popularity, along with Pininfarina's 50th anniversary in business, the Italian company built a run of 1000 Spiders, painted in champagne silver and decked out with brown or cream leather upholstery, electric windows and Abarth-style 5.5J alloy wheels. This model can be spotted by the special edition number below the Pininfarina badge, and on the dash and key fob.

Yet behind the celebrations, and despite the fact that in North America it had the open sports car market almost to itself, the Spider's success was beginning to falter. Even more significantly, Fiat was now keen to pull out of North America altogether, having never really grasped the realities of running a network of agents in a country of such size and with the world's most demanding consumer society. Fiat's reputation for producing rusty cars with uncertain electrics was at an all-time low, with beleaguered American owners joking that Fiat stood for 'Fix It Again Tony'.

In a drastic move Fiat took the decision to bail out of the sports car market altogether. The Spider's volumes were low by company standards and its production took up valuable saloon car capacity, while the additional emissions equipment that had to be fitted to sell in America cut profit margins to the bone. It is even said that the company, still recovering from the strike-ridden late 1970s, was losing money on every Spider sold.

Pininfarina takes over

Thus, towards the end of 1981, Fiat handed over the whole production process to Pininfarina. Whereas the coachbuilder had previously sent completed bodies to Fiat for the running gear to be fitted, Pininfarina would now be building, and marketing, the whole car itself, buying in the mechanical parts from Fiat.

Between March and July 1982, production of Fiat and Pininfarina Spiders overlapped as Fiat finished off the final 2000 Spiders and Pininfarina built its first own-brand cars. For the record, Pininfarina cars were given the chassis number prefix 124CS, Fiat models 124DS.

At this time the Spider was one of seven different bodies being built by Pininfarina, the inventory ranging from exotica like the Ferrari 400i, Lancia Gamma and Peugeot 504 coupés, right down to the comparatively humble Talbot Samba Convertible.

ESSENTIAL FIAT 124 SPIDER & COUPÉ

Externally little changed for the post-1981 Pininfarina Spidereuropa apart from a slightly more purposeful appearance and door mirrors mounted on the quarter-light glasses, but the cabin became a little more inviting, helped by the redesigned centre console. On nose and tail, new Pininfarina badges ('F' is for Farina) replaced Fiat motifs in the circular panel indents.

UNITED SPIDERS OF AMERICA

Compared with earlier Spiders, the tail lights are now bigger still. The 2000 engine, fuel-injected from 1980, took mainstream production through to the finish.

A change like this was deemed worthy of a change of name, especially as it marked the reintroduction of the model to the European market, now that American sales had slowed down. In Europe the 2000 injection model was now to be known as the Pininfarina Spidereuropa, with all Fiat badges banished. Outwardly it now sported a longitudinal rear number plate and remote-controlled side mirrors mounted on the quarter-lights, while inside the seats had changed to what looked like a curious towelling trim. This was actually a seat cover, intended to put an air gap between the driver and the vinyl or leather of the seat itself. The rear seat, always next to useless, was now abolished and replaced by an upholstered shelf.

In the States the name changed to the Pininfarina Spider Azzura and the car was to be imported and marketed by Malcolm Bricklin's International Automobile Importers Ltd. The Azzura (it stands for blue in Italian, although the car was offered in a complete range of colours) was a no-options package, positioned somewhat upmarket from the Fiat-badged car in an

35

The Turin Show of 1982 saw a supercharged version of the Spider launched on the Pininfarina stand, where a Peugeot 505 estate was a reminder of the company's more humble design and manufacturing activities. Power output was 135bhp from the Lancia-engineered Volumex, a title that was placed on the bonnet of the production model. Wider Pirelli tyres, Speedline 'eight-hole' alloy wheels and plastic wheel arch extensions were other developments.

attempt to increase profit margins. The base price now rose $16,995, compared to $12,290 the previous year for the Fiat-badged 2000 Spider.

In an attempt to justify this increase the car came with leather trim, a good stereo/cassette and electric windows as standard, and was generally better finished, with a more neatly trimmed dash and luggage area and a tighter-fitting top. But mechanically it was the same old story. Both Pininfarina Spiders were still fitted with identical 1995cc engines, rated at 102bhp in US catalysed form and 105bhp in Europe without a catalyst. Sadly the car was never fitted with the 122bhp twin-cam engine used in other 2-litre Fiats and Lancias of this period.

Instead, in a bid to give the Spider a sportier image in its declining years, Pininfarina struck a deal with Lancia to supply its supercharged twin-cam engine. The result was the Spidereuropa Volumex (VX for short), a car second only to the Abarth Stradale of the 1970s in the Spider desirability stakes and first shown in 1982 at the Turin Show, with Abarth badges and the matt red finish used on the 16-valve Abarth CSA in 1975. The supercharger technology was lifted from Lancia's Beta range, where its first production application had been in the unloved Trevi saloon.

Why supercharging? Pininfarina's answer was that it was cheaper to produce and did not require such extensive re-engineering of the engine as a turbo. It also pointed out there was no lag with a supercharger, which runs a constant low level of boost through its compressor, on account of it being mechanically driven from the crankshaft, forcing the fuel/air mixture into the combustion chambers at above atmospheric pressure and working consistently throughout the rev range. With a turbo there was usually a delay, particularly with early

Yet another wheel variety – an eight-spoke design from Speedline – came for the 1985 model year, the last for the Pininfarina Spidereuropa.

systems, while the rotors ran up to 100,000rpm, hence the term 'turbo lag'. Supercharged engines, with flatter torque curves, give good low-speed pick-up just at the point where a turbo might feel 'flat'. Historically supercharged cars tended to be thirstier, but Pininfarina claimed the VX would have fuel consumption figures consistent with a comparably quick, normally aspirated 2.7-litre car.

The Roots-type compressor was crankshaft driven via a toothed belt and two pinions to synchronise the rotation of the twin-lobed rotors. These turned in opposite directions within a two-ported housing shaped like a figure of eight, with one side of the unit connected to a modified Weber carburettor fed by a mechanical pump and the other to the intake manifold. The compressor had a positive displacement of 1130cc, giving a maximum boost of 0.4 bar or 5.7psi, with a release to limit it. Power output was raised to a healthy 135bhp at 5500rpm, with maximum torque of 152lb ft at 3000rpm.

The twin-cam became rather exotic in some of its details to deal with the added heat that was generated. Compression ratio was dropped from 8.2:1 to 7.5:1, valve timings and electronic ignition were tweaked, exhaust valves were now sodium-cooled and the head gaskets were ceramic. In the heart of the engine there were new piston crowns, with chromed rings for scraping oil from the barrels. Underneath was an Abarth exhaust system.

Naturally there were also changes to the drive train. New castor angles, anti-roll bars at both ends and Bilstein gas-filled dampers tightened up the handling, while ventilated discs and a better servo sharpened the brakes. The ratios in the five-speed 'box were closed up and the clutch beefed to handle the new-found torque.

Outwardly it was impossible to miss the eight-hole Speedline split-rim wheels, fitted with 195/50 VR 15 Pirelli tyres that filled the arches in a much more aesthetically pleasing way by leaving less space at the top. The wider rubber was further accommodated by adding black plastic trims to the wheel arches, the front ones merging neatly with the new, deeper, chin spoiler. Exterior embellishment included the inevitable coachline and spotlights, while inside leather, tinted glass, electric windows and a remote boot catch where all standard. Although the driver was presented with much the same layout as the standard model, there was a boost gauge for the supercharger, while the cockpit was finished off with a handsome three-spoke Abarth steering wheel sporting a chunky leather-covered rim.

There was, of course, a price. At 22,000,000 lira, the VX was nearly a third more expensive than the standard Spidereuropa. But at last the ageing car had real muscle. Top speed leaped to 120mph and 60mph was reached in a forceful 8.5sec. Yet these bald figures do not convey the full difference in flavour. The VX, as easy to start as the normally aspirated Spider, and hardly any noisier, would pull from as little as 600rpm in third of fourth gear without complaint, running right up to 6000rpm without gaps or flat spots – just one seamless rush of energy, almost as if it had a V8 hiding under the bonnet.

Writing in *Car* magazine in November 1982 about a VX prototype that was still wearing Abarth badges, Ian Fraser also praised the car's stiffer suspension. 'The Abarth has about as much roll as your average office desk and a grip like a bulldog on a burglar's leg,' he enthused dramatically. The state of the art Pirelli P7s meant there was simply no question of breaking adhesion in the dry. Push hard in a tight corner and the VX would reluctantly understeer, while faster, longer curves only provoked well-mannered neutrality.

Yet the car was not a commercial success. Only 500 had been built, mostly in metallic red (although there were 42 black ones), before Pininfarina finally pulled the plug on all Spider production in 1985 to make way for the Cadillac Allante on its production lines.

In their final run-out year both versions at last

Still a fine car, but the original Spider's purity of shape had become fussily embellished in the Pininfarina Spidereuropa phase. This high view shows the flat luggage platform that replaced a conventional rear seat in 1981.

received rack and pinion steering, with 3.5 turns lock-to-lock, while the injected Spidereuropa was fitted with front discs increased in diameter from 8.94in to 10.83in, along with bigger calipers. The final Spidereuropas also had a new style of eight-spoke Speedline alloy wheel, still Pirelli P6 shod, with a revamped cooling system, new steering wheel and reading lamp in the glove box.

A few late Spiders were converted to right-hand drive and sold as personal imports in the UK by Basingstoke Nissan dealer, Dennis Hands, who had ordered 150 from Pininfarina in July 1984. There were few takers at £9000 each and only seven were sold. In any case, by then Pininfarina was already beginning to oust the Spidereuropa from the production lines. It had been a relatively low-volume model from the beginning, with a total output between 1982-85 of only 9400. Production had peaked at 2500 in 1983, while only 186 of the 1200 built in the last year were exported to America.

The last Spidereuropa came off the production line at 3pm on 30 July 1985, although the final cars did not all find homes until November.

Unless you count the Bertone-built Strada Abarth, it was to be another decade before Fiat offered a purpose-built sports car with the launch of the Barchetta in 1995. In some ways this car has its appeal pitched between the 124 Spider and the X1/9. A front-wheel drive machine, built on the Punto floorpan and styled in-house by Fiat, it is bigger than the X1/9, more aggressively sporting than 124 Spider, tops 120mph with ease and has been well reviewed around the world.

For the British, though, it shares one depressing fact with the 124 Spider – Fiat will not be building any with right-hand drive.

ABARTH

The 124 Abarth Rallye, or *Stradale*, a homologation special built in small numbers to qualify the works cars for international rallying, has almost mythical status in Fiat circles. It might not have been stunningly successful in competitions, but it has pedigree, rarity, exotic fitments and strong performance, although the supercharged VX is marginally quicker and rarer.

To qualify a car as a standard production model for the Group 4 category, a manufacturer had to build 500 units. And although the Spider had proved strong and reliable, the works drivers asked for a model that was not just lighter and more powerful, but had independent rear suspension.

The first 500 Abarth Spiders – designated CSA – were

The icon among Fiat 124 Spiders. Only 1013 examples of the sparkling Abarth version were built and today replicas abound – but this is a real one. Cover from *Motor* shows that production versions did not come only in the familiar red.

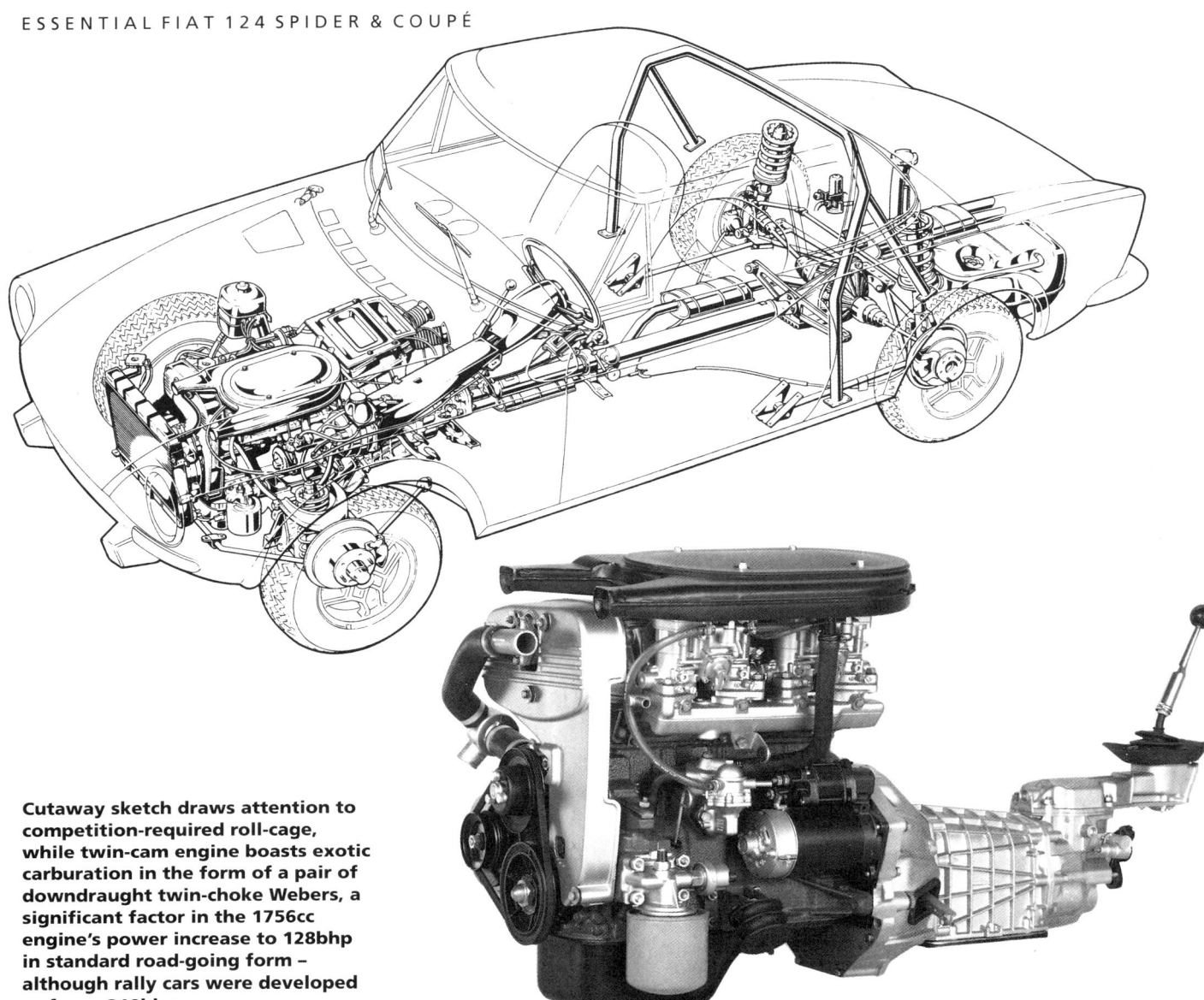

Cutaway sketch draws attention to competition-required roll-cage, while twin-cam engine boasts exotic carburation in the form of a pair of downdraught twin-choke Webers, a significant factor in the 1756cc engine's power increase to 128bhp in standard road-going form – although rally cars were developed as far as 210bhp.

built between September 1972 and January 1973 for the home market (although Fiat claimed it had completed them by November 1972). Another 513 were then made by September 1974, when they became eligible for homologation into Group 3 for production GTs.

The key to Fiat's involvement in rallying, and the company's ambitions in the World Rally Championship, was the acquisition of Abarth in August 1971. From then on this company worked exclusively for Fiat, although the two's fortunes had long been inextricably linked.

Abarth was named after its founder, Karl Abarth, a motorcycle racer and engineer who was brought up in Vienna. He moved to Italy at the end of the Second World War, aged 37, and changed his name to Carlo. It was not long before he came to prominence when Ferry Porsche invited him, along with Rudolf Huruska (later of Alfa Romeo), to act as his racing representative in Italy. Abarth's first job was to develop an ill-starred supercharged flat-12 Formula 1 car for Piero Dusio of Cisitalia, but the project foundered when the cost of items like four-wheel drive and variable torque-split became more than even Dusio could afford.

Undeterred, Abarth, armed with one D46 single-seater, two 204 Sports roadsters and three rolling chassis as severance pay, began producing sports exhaust systems for a variety of cars as a way of funding his own racing ambitions. Establishing his workshop in 1949 at 38 Corso Marche, Turin, he built his first complete car the next

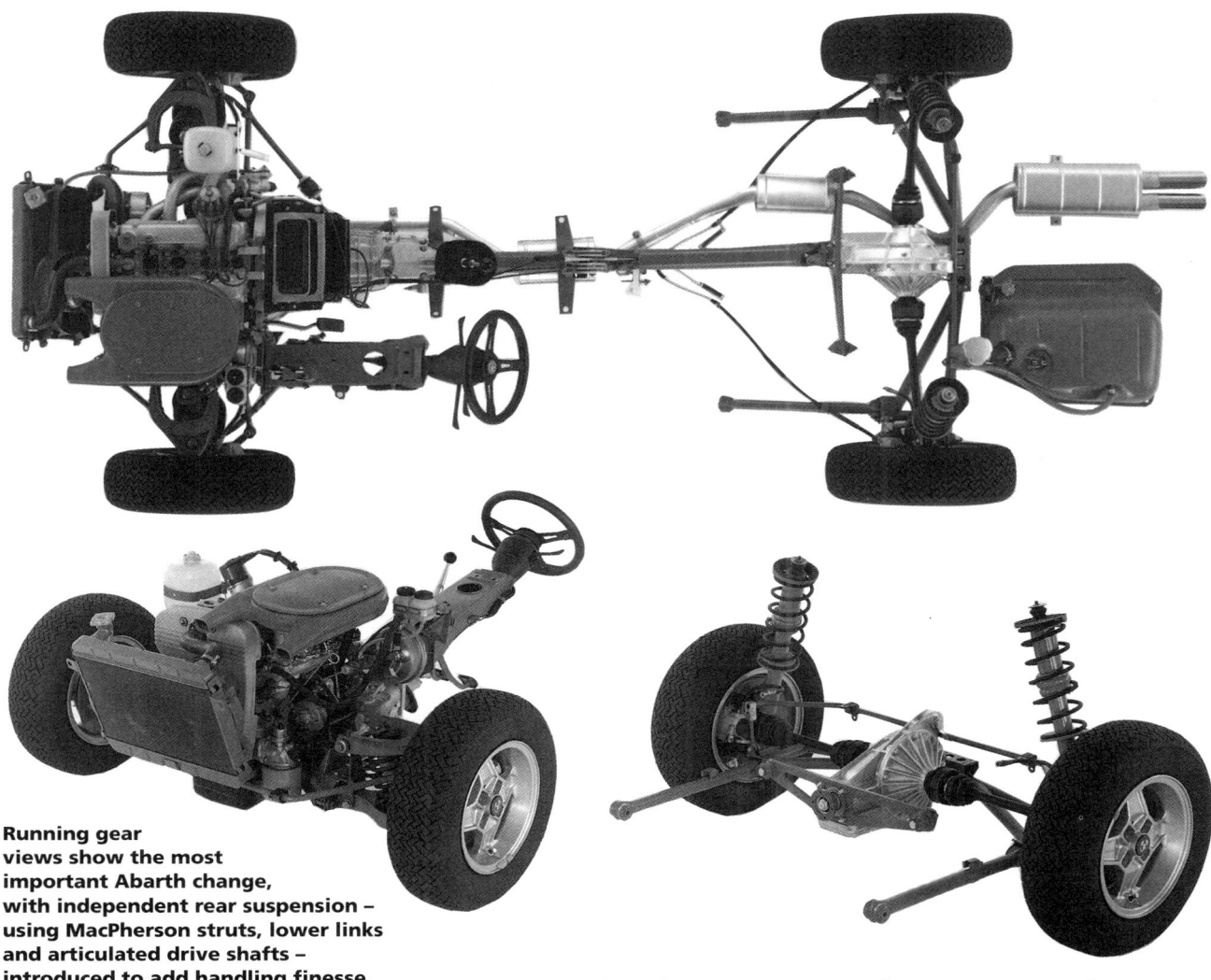

Running gear views show the most important Abarth change, with independent rear suspension – using MacPherson struts, lower links and articulated drive shafts – introduced to add handling finesse. Pirelli CN36 tyres on lightweight magnesium Cromodora wheels were other features.

year, immediately showing where his thrust lay. Although the Fiat 205 Berlinetta, with bodywork by Vignale, only had a 1090cc engine, it was capable of 114mph.

By the late 1950s Abarth was offering more elegant Vignale-styled coupés with radically tweaked Fiat six-cylinder engines. Alongside he had also produced the 850 TC, based on the humble Fiat 600; it became famous, both as a race winner and for its distinguishing feature of having the engine cover permanently propped open to stop the machinery overheating. Next came the radical *Bialbero* models, producing 80bhp from 1000cc and equipped with five-speed gearboxes, that became a familiar and dramatic sight in touring car racing and gave him the manufacturer's championship title between 1962 and 1966. In some ways the company was similar to Cooper in England, but much bigger, employing 350 workers by the mid-1960s and producing over 3000 cars in 1964 alone.

The family tree of Abarth models is a complex labyrinth of semi-production cars and pseudo-racers, not all of which were Fiats. Alfa Romeo 2000s were 'Abarthised' in the early 1950s and the Porsche-based Abarth Carrera won its class at Le Mans in 1960, while in 1963-64 the company built its own versions of the Simca 1300 and 1500, clad in elegant coupé bodywork. Less elegant, but a real 'wolf in sheep's clothing' that had huge appeal for boy racers, was the Abarth-Fiat 595, based on the tiny 500 and capable of 90mph.

Another string to the Abarth bow, which proved to be a valuable publicity tool, was a sequence of cars built to break speed records. These svelte single-seaters, with all-enveloping streamlined bodywork by Pininfarina, notched

Massive twin-intake air filter conceals Weber carburettors, but special four-branch exhaust manifold is visible. Cabin on this example is stripped out, but cars came from the factory with full trim, including carpet on the floor – correct aluminium dashboard finish adds purpose.

up a string of successes, starting in 1956 and lasting right through the 1960s, and held records in classes for 500, 750, 1000 and 1100cc.

Soon after selling out to Fiat in 1971, Carlo Abarth retired, finally dying in 1979, by which time his factory had become the headquarters for the Lancia rally team. Meanwhile the significance of the famous Scorpion badge – his birth sign – gradually eroded until today it has been reduced to providing stylish, but superfluous, 'look-faster' bits for everyday Fiats. Yet his legacy lives on in some of the most wickedly unforgettable road/race machinery ever built, along with a huge selection of bolt-on hardware that inspired much of today's tuning industry throughout Europe.

The 124 was one of the last cars with genuine Abarth provenance, conceived in the twilight days of Carlo's tenure. In 1971, a year before the 124 Abarth Rallye was homologated, Abarth badges had already begun appearing on Spiders equipped with 1608cc engines, Colotti five-speed gearboxes and uprated BS live axle rear suspension. Production of the proper 124 Abarth Rallye started in September 1972, but it was not until the Turin Show in November that Fiat finally came clean and revealed the model on the Pininfarina stand.

The manufacture of the car was a complicated process that involved much toing and froing. Bodyshells were taken at random off the Spider production line at Pininfarina and sent to the Abarth factory, where they were clad in lightweight panels, then sent back. Meanwhile engines arrived at Abarth from Fiat at Lingotto, eight kilometres away, to be suitably modified and returned, along with new suspension parts. Pininfarina sent the body back to Fiat for final assembly on a special production line, the finished car went back to Abarth for testing, and was then finally returned to Fiat for quality checks. With all this going on it was little wonder the cars were expensive – £2250, against £1500 for a standard 1800 CS model. Furthermore, they were only available in red, sky blue or white, with matt black bonnets and boot lids.

Curiously, the Dutch highway police tested three, with hard-tops containing a removable targa section, as

Period view of Abarth *Stradale* interior shows handsomely sporting bucket seats, chunky steering wheel and simplified door trim.

potential patrol cars. But for many 124 CSA owners there was only one thing to do – buy all the choice bits off Abarth and make themselves a proper rally car. Fiat had produced a dream machine for rich boy racers.

Although Abarth concentrated all its attention on making the shell lighter, by the time the car had various bits and pieces added the reduction in weight over standard was only 48lb. The paring process included a glass-fibre boot lid and bonnet, double-skinned at the edges and retained by rubber straps, and aluminium skins for doors, sill valances and scuttle panel. Holes were even drilled in the door hinges and the standard bonnet prop replaced by such a rudimentary version that it barely supported the weight of the panel. In a final flourish steel was cut away from the quarter panels behind the seats.

The detachable hard-top, an option on the standard Spider, was manufactured from glass-fibre and given a rear window made from Perspex. It was then attached to the top rail of the windscreen using clasps normally provided for the soft-top, before being bolted to the back scuttle panel. There was no provision for a soft-top. The standard bumpers were replaced by rubber block overriders, while the glass-fibre wheel arches allowed Fiat to fit Cromodora wheels as wide as 8in. Extra inner air intakes at bumper level assisted the oil cooler, while outer ones directed air onto the brakes and shock absorbers, which were unchanged. A more accessible fuel cap, with a screw fit, was mounted behind the back window on the left.

The cabin was stripped to bare essentials and equipped with an Abarth steering wheel and bucket seats containing built-in headrests, with the option of corduroy-trimmed Recaros. On the dashboard the wood veneer disappeared in favour of stark and functional aluminium, while the centre console and glove box were thrown out to make way for Halda navigation equipment and map-reading lights. Even the door trim was simplified to a black plastic card, without armrests, while the only real concession to civilisation was retaining the floor carpet. Meanwhile the shell was further braced by a roll-over bar, bolted to the floorpan forward and aft of the rear wheel arch housings.

Although the essentials of the front suspension remained, there were different uprights, a rose-jointed anti-roll bar and additional reaction rods attached to the lower part of the struts. At the back arrived the independent set-up the drivers had requested, and which both improved grip and handling, along with reducing the unsprung weight inherent in a live rear axle.

Springing was by MacPherson struts, with the combined spring/damper bolted to the top of the upright, while the lower link, an inverted wishbone, had its inner end attached to a bracket behind the differential

Staged publicity shots varied from the contrived (uninvolved female crew in spotless car with hard-top removed) to the believable (dirt-streaked machine tearing through a corner).

casing. A longitudinal torque arm, with rubber bushes, ran from the outer end of each wishbone, using the same pick-up points as the standard Spider. This set-up, easy to adjust, was to see further service in the Spider's more successful competition replacement – the 131 Abarth.

The alloy differential housing was attached to the underframe by two rubber-bushed supporting struts where it met the propshaft, with another strut behind the crown wheel and pinion, and the drive shafts were fitted with CV joints. The elegant four-bolted wheels were 5.5J × 13 CD30 magnesium Cromodoras, carrying Pirelli CN36 185/70 VRs. Today's Spider owners can buy copies, but made from less exotic, and heavier, aluminium.

A host of homologation tweaks was provided for the competition-minded owner – special halfshafts, 68-litre fuel tank and guard, external oil cooler, uprated dampers front and rear, lower trailing arm suspension mountings with spherical joints, sump guard, dual brake servo, stronger front cross-member and straight-through exhaust system. For those mindful of accidents the roll-over bar could be extended to a full cage by adding bars along the roof and down each side of the windscreen to floor level.

The 1756cc engine, although expensively blueprinted, was surprisingly stock. At a glance the most obvious changes were the distributor drive being taken off the exhaust camshaft, an improved four-branch exhaust manifold and twin downdraught 44 IDF Weber carburettors. Inside there was more valve overlap on the cams, although the pistons were identical to the 1608cc Sport. This starting kit gave 128bhp at 6200rpm.

But a host of homologated parts could then be added – high-lift cams, nitrided crank, hardened flywheel, ventilated crankcase, high-compression pistons, nine-blade cooling fan, improved four-into-one inlet manifold, carburettor filters and trumpets. The list went on, while power grew to anything between 170 and 190bhp. Meanwhile there was a range of optional ratios for the five-speed gearbox, which was otherwise standard.

Sting in the tail..................................

Driving one of the few *Stradale* CSAs still in road trim, it feels surprisingly tame and friendly. Noisier, harder riding and more nimble, certainly. Yet not that much quicker and perfectly acceptable everyday transport – apart from the rudimentary bumpers, delicate body panels and, of course, the glorious noise.

The first thing you notice is the lighter doors, while the body-hugging optional Recaros and stripped-down cabin give a sense of purpose the standard car lacks. You grip a lovely thick, small sports wheel, with a leather rim and Abarth's Scorpion emblem on the horn push, while the other stalk controls for lights and wipers are standard.

As you build up speed the lack of refinement soon hits you – hardly surprising when there is practically no sound deadening and the differential is bolted straight to the floor a foot or two behind you. There is also plenty of wind noise, as the sealing is far from perfect, while the crashy ride makes the glass-fibre hard-top rattle and shake.

Despite its modifications, a standard Abarth engine remains utterly docile, making up for what it lacks in urgency by the noise from its big-bore twin tail-pipes. The steering is tiresomely heavy at low speeds and the brakes a little over-boosted – rather like the ordinary Spider – although the throttle pedal allows fine control and gives instant response. But you still have to wind the

Noisy, harsh and firm – on the road the Abarth version adds sporting edge. Apart from engine tune, improved performance results from lighter construction, exemplified by exposed glass-fibre weave on bonnet underside.

engine hard to feel any real response. The 0-60mph time, in the order of 9sec, needs two gear changes and is not much better than the standard car, despite the respectable power to weight ratio of 139bhp per ton. Yet a healthy CSA still hits its 7000rpm red line more eagerly than a standard Spider, helped by the close ratios in the slick

Ultimate 124 Spider looks purposeful from any angle, and Abarth motifs are liberally applied – on wheel centres, front wings and bonnet catches.

gearbox, which is a joy to punch round its well-defined gate, while the clutch is light and fairly sharp, but in no way difficult. Top is geared to just 17.2mph per 1000rpm, so you can forget relaxed, or frugal, motorway cruising.

The suspension is rock-hard, unforgiving, and easily upset by bumps and ridges. In return it delivers faithful, neutral handling, with steering full of feel and wonderfully direct as it wriggles over every bump. Once again, as with all Spiders, it is astonishing to think this is achieved through a humble worm and roller box. The car corners as flat as a pancake, with copious grip, even on period Pirelli tyres, and as the g-forces build up you bless the bucket seats, while you marvel at the eager turn-in and the compact and solid feel of the whole car.

That Fiat was prepared to build and homologate for Group 4 showed how much faith it had in the car. It is easy to see how the company thought it might have a winner, quite apart from its intention of building an image of strength and reliability. It was therefore doubly unfortunate the Abarth had hardly moved into its stride before it was overtaken by the all-conquering Lancia Stratos.

For enthusiasts, though, not only did Abarth's work contribute to improvements to the standard car, but it gave them the chance to buy something very special. Not surprisingly, the 124 CSA Stradale Abarth Spider, to give it its full title, has inspired countless replicas. But if you insist on the real thing, you will not only have to spend a long time looking, but ensure beforehand you have an exceptionally deep pocket.

RALLY TIME

The 124 Spider's rallying career may have been largely a succession of false dawns, but the best result was truly sensational – works cars finished 1-2-3 on the 1974 TAP Rally in Portugal. As third-placed Markku Alen opposite-locks, spectators turn their backs against the dirt raised from the rear wheels.

Fiat as a company was very new to any form of competitive activity when it began supplying clandestine technical support to private owners rallying their Spiders.

Despite the independent exploits of Carlo Abarth with Fiats, it had been more than 50 years since the company itself had retired from Grand Prix racing in the 1920s, leaving the field to its more sporting rivals, Alfa Romeo and Lancia, while it concentrated on the more mundane, if lucrative, business of supplying cars to the Great Italian Public.

But now that both Lancia and Ferrari had been absorbed into the organisation (both were acquired in 1969), the company culture had begun to change. Competition, with its attendant publicity, was looking an attractive way of promoting core products. And in the Spider Fiat saw the vehicle to take it back into the arena.

Whether this was wise was debatable. Either way, the Spider's timing was never right: although full of promise, it was destined to be one of the 'nearly' cars of international rallying. In its first guise it was too heavy and under-powered to outpace the lightweight Alpine-Renaults and Ford Escorts. Then, just as it was coming right, it was outgunned by the purpose-built in-house opposition of the wickedly fast Lancia Stratos.

Nonetheless, apart from introducing a welcome new

After a discouraging start in 1970, the Spider broke the ice on the 1971 Monte, where Hakan Lindberg – the first of many rallying stars hired by Fiat – finished seventh, paired with Solve Andreasson.

element into the sport, the Spider had its moments. It had the strength, speed and staying power to win the Polish Rally in 1973, the Portuguese TAP in 1974 and 1975 and, behind the works Stratos, chalk up an impressive 2-3-4 in the 1975 Monte Carlo Rally as a fitting swan song.

More importantly for Fiat, its short career paved the way for its highly successful competition successor. Not only did the Spider enable Fiat to bed down running a works team, but the car was used to develop the drive train for the 131 Abarth that was to really put Fiat back on the map by winning the World Rally Championship three times.

The Spider was a contender for the World Championship of Makes between 1972 and 1975, coming second to Lancia each year except 1973, when the title went to Alpine-Renault. Even before the definitive Abarth CSA was homologated, the works drivers Raffaele Pinto and Gino Macaluso won the 1972 European

Drivers' Championship in a 1600 BS, and in its time the Abarth was driven by the top Scandinavians Bjorn Waldegaard, Rauno Aaltonen and Markku Alen.

The attraction of the Spider was that it was strong and rugged, an ideal size and had enormous tuning potential. In addition it was ideally suited to the new breed of loose surface special stage rallies that were developing as the sport was driven off tarmac roads.

Fiat started clandestinely, unofficially hiring the burly Swedish rally star, Hakan Lindberg, as their top driver, and doing a deal with Pirelli to supply tyres. The company then entered not just Spiders, but 125 saloons, in the names of their drivers, although few people were fooled by these 'unofficial official' cars, when they carried Turin registration plates and had service crews sporting Fiat overalls. At the same time Fiat carefully combed events to choose those to which they thought the cars most suited and which had sufficiently high profile to gain maximum publicity, while hopefully having the weakest opposition.

But the Spiders' early performances were patchy. In 1970 all four of the AS models entered in the Monte Carlo Rally retired, as did the BS of Alcide Paganelli and Domenico Russo in the British RAC, after hitting a tree.

The 1972 season saw Fiat come out of the closet with its works role, now based at Abarth. The first event, the Monte, brought eighth place for Raffaele Pinto and Helmut Eisendle in a regular Spider (left), while a couple of months later in Portugal an Abarth version (below) ran as a prototype for the first – and only – time that year, driven to an encouraging fifth place by Alcide Paganelli and Domenico Russo; bonnet vents and lack of bumpers distinguish it.

At a lower level, though, there was a glimmer of what might be to come when Paganelli/Russo won the Italian Drivers' Championship.

In 1971 matters improved. Lindberg finished seventh on the Monte, with Paganelli/Russo 24th. Then the same duo came second in the Austrian Alpine, while Pino Ceccato and Helmut Eisendle gained fourth in the Acropolis. The Portuguese TAP was not so successful. All three 'unofficial official' cars retired with rear suspension failure, and on the RAC, along with a solitary 125 saloon, they were also-rans behind Saab, Porsche, Ford and Lancia.

But now Abarth, absorbed by Fiat in August, began to make its mark. The Fiat works connection was officially declared for the 1972 season and an open assault on rallying honours declared. The competition department moved over to the Abarth works, where it was overseen by Giovanni Sguazzini, the director of the technical department of Fiat's passenger car division, now Carlo Abarth himself had retired. All the resources of Fiat were made available, while in turn Fiat's engineers used the rally Spiders for testing and development, learning valuable lessons they later applied to standard production cars.

The season started with a much modified Spider, now putting out around 165bhp and fitted with a Colotti T70 straight-cut gearbox. Driven by Pinto/Eisendle, it came eighth in the Monte, then gained its first European Championship success in February, with a win in the Costa Brava. In April four Spiders were in the first seven places in the Isle of Elba, then Pinto/Macaluso won the Polish. In May Lindberg/Eisendle won the dusty, hot Acropolis, with another Spider driven by Smania/Zanuccoli seventh, and Lindberg/Eisendle followed through by winning the Austrian Alpine. Spiders now lay third in the World Rally Championship.

The Portuguese TAP provided the inaugural, and only appearance that year, of the CSA Abarth, run in the

Fiat's 'home' event was the San Remo, where 1972 produced the best team effort of the season. Hakan Lindberg and Lars-Erik Carlstrom headed the 3-4-5 result, despite their Spider's Colotti gearbox having to be changed – a task that these three mechanics completed in a feverish 20 minutes.

prototype class as it had yet to be homologated. Paganelli/Russo took fifth place in a fascinating event that showed how much diversity there was on the scene at that time. The first six cars were all different makes – BMW 2002 Ti, Alpine-Renault, Citroën SM, Datsun 240Z, the Abarth and Porsche 911S – but what cheered Fiat was the Abarth winning five stages outright. It might even have been a serious contender for overall victory but for fuel pump trouble. Not so cheering was another Spider, driven by Waldegaard, leaving the road and careering 180 yards down a mountain.

Giving best to the works Lancia Fulvias, the Spider of Lindberg and Lars-Erik Carlstrom then took third place on the San Remo, with Giulio Bisulli and Arturo Zanuccoli fourth, and Luciano Trombotto and Giuseppe Zanchetti fifth, despite brake fade and transmission problems. But none featured among the front runners in the RAC.

By January 1973 the Abarth Rally CSA, with its independent rear suspension and multiple modifications, had been homologated after Fiat claimed it had built the necessary 500 by November, only four months after launch – a claim which, though accepted, should be taken with a large pinch of salt.

The competition cars were manufactured in a similarly tortuous way to the *Stradale* Abarths. Cylinder heads, transmissions and rear suspension components were manufactured in the Abarth factory, then taken to the Fiat factory at Lingotto to be mated with bodyshells which had meanwhile come in from Pininfarina. The cars then went back Abarth to be so comprehensively worked on that they ended up, after appropriate strengthening, 22kg

Fashion-aware Alcide Paganelli – notice the bright red flares! – was not part of the successful three-car team result on the 1972 San Remo, but looks confident before the event.

heavier than standard. But, with up to 176bhp, performance was not only electrifying, but brutally uncompromising.

The Colotti gearbox, with its shrieking straight-cut gears without synchromesh, needed extremely accurate double-declutching and the higher-geared steering was enormously heavy, although the brakes, with no servo and competition pads, were an improvement on the over-light standard items. The cars were equipped with close ratios and short sprint gearing that allowed the drivers to make the most of the top-end muscle, with power coming in hard and fast as the rev counter swept past 7000. But what really made the car such a favourite, with both drivers and spectators, was its classic rear-drive handling, that necessitated hanging out the back end in glorious power oversteer.

The works cars featured batteries mounted in the boot and a flat-bottomed fuel tank, although the interiors looked much like any *Stradale*, except that all the switches were grouped on a single panel. Outwardly the spats on the wheel arches had grown to accommodate fatter 8in rubber, while the standard livery was a distinctive red, with a wide white stripe along both sides and hard-tops in either white or black.

For 1973 Fiat amassed a team of 34 mechanics to prepare the cars and planned to start the season with a total of 28 cars – 14 for recce work and 14 for the actual events. Now the car was fully homologated, it was deemed time for a serious attempt on the inaugural World Rally Championship, with the chief adversary the glass-fibre Alpine-Renault A110, developing 175bhp and weighing in at just 1500lb, with a reputation for being not just fast, but reliable.

The Monte was not an auspicious start. Waldegaard/Russo retired early, while Paganelli hit a wall when a gear selector failed. But Pinto/Bernacchini, driving to finish, came a creditable seventh. There was better news on the Swedish, when Lindberg and Solve Andreasson set a succession of best stage times and finished fifth.

But there was trouble in Portugal after Paganelli left the road and knocked down a TV cameraman, while Waldegaard only managed sixth on the Moroccan after the suspension collapsed. The Polish went better, Achim

With the Abarth homologated, the 1973 season began on the Monte and looked promising, but seventh-placed Raffaele Pinto and Arnaldo Bernacchini were the only works finishers.

Warmbold winning after Therier's A110 Alpine was disqualified for missing a stage. Yet it was really a hollow victory. The Alpine was the only other works car in the event, mostly a ragbag of local Wartburgs and Polski Fiats.

Warmbold distinguished himself more by taking eighth on the 1000 Lakes, an impressive performance considering he had never driven in Finland before, while Rauno Aaltonen, who had now joined the team as a 'freelance', took a good second on the Acropolis, with Lindberg and Arne Hertz fourth.

For the San Remo, on home territory of Northern Italy, no fewer than five factory cars were entered, taking second, fourth, fifth and sixth, but conceding victory to the Alpine-Renault, while the RAC was the usual disaster, with Maurizio Verini careering off into the undergrowth on an early stage.

For 1974, the cars were homologated into Group 3, for production Grand Touring cars, and fitted with 16-valve Abarth 236/B heads. Engine capacity was raised to 1839cc, adding another 30bhp to achieve 210bhp. By the time the season ended capacity had crept up to 1850cc, with twin downdraught 44 Weber carburettors, and a ZF 'box replacing the Colotti. Pairs of spotlights were incorporated into the bonnet and grille, with further revisions to the instrument panel, transmission, suspension and brakes. In an effort to cut the weight problem, the body panels and hard-tops became even thinner, and it was even said the cars were given fewer coats of paint. Colours, too, had changed – the works cars now attired in matt red with yellow wheel arches, the semi-works Jolly Club Team cars white with red trim, and the Olio-sponsored vehicles dark blue with contrasting yellow. There were driver changes as well. Waldegaard and Aaltonen had moved on and Scandinavian Markku Alen was now signed as a new rising star.

The season started triumphantly. Spiders were first, second and third on the TAP Rally of Portugal, the cars handled respectively by Pinto/Bernacchini, Paganelli/Russo and Alen/Kivimaki. But after that there were to be no more outright victories, although Alen took third on the 1000 Lakes and Robin Ulyate was tenth on the gruelling African Safari, the first time Fiat had attempted it.

Signs of trouble in 1973. With help from *gendarmes* and spectators, Raffaele Pinto (above) got out of this snow drift high on the Col de Turini to finish seventh on the Monte, while Maurizio Verini (left) crunched a rear corner during his second-place contribution to the Fiat team's excellent 2-4-5-6 result on the San Remo.

The battle was now between the Spiders and Lancias of Stratos and Beta Coupé ilks, with the latter using a similar 16-valve twin-cam engine. Although Bisulli took second, the home ground San Remo was a horror. No fewer than four works cars were eliminated on the first special stage – one ending up hanging precariously over the parapet of a bridge, another wrecking its rear suspension, while the fourth collided with the third after it had spun on a hairpin. Sergio Barbasio's car then put a rod through the block, while Alen, who had been lying second, cracked a rear wishbone.

Attention was then redirected to North America where, to boost sales, three works cars were entered in the Rideau Lakes Rally in Canada. All three promptly retired ignominiously on the first leg. There was some consolation in the American Press-on-Regardless which followed and Alen would actually have won, had not Fiat been successfully accused of car-swapping in the *parc fermé*

Two decent results in 1974. Markku Alen (right) finished second on the Press-on-Regardless Rally in Michigan and would have won but for a rules infringement, while Leo Kinnunen (below) came 14th on the RAC in a specialised forest-stage environment that did not suit the Fiats.

– the officials relegated him to second behind the Renault 17 Gordini of Therier/Delferrier.

The RAC went a bit better than usual, with Aaltonen 12th and Kinnunen 14th in the face of specialist forest stage opposition from Saab, Ford and Lancia. But the final round – the Tour de Corse – proved a disaster. A wrong tyre choice sent most of the works cars careering off the road, with only Fulvio Bacchelli and Bruno Scabini surviving to take sixth behind four Alpine-Renaults and the winning Stratos.

In the end Fiat was second behind Lancia in the World Championship. There was some domestic consolation in winning the Italian Rally Championship, with some more useful publicity when Tominz/Mamolo became European and Italian Women's Champions.

For 1975, with Abarth production over, the car was rehomologated once again, this time into Group 4, where it became a testbed for the 131 Abarth. The CSA was now putting out 210bhp at 8000rpm, with a prototype 16-valve head and Kügelfischer fuel injection. Outwardly it had become even more radical, with larger glass-fibre wheel arch extensions, air vents behind the front wheels and air scoops ahead of the rear wheels.

Although it was the Lancia Stratos of Sandro Munari that set the pace on the Monte, the two Finns, Hannu Mikkola (with Jean Todt) and Markku Alen (with Ilkka Kivimaki), finished second and third, followed by the Italian pairing of Bacchelli/Scabini in fourth place.

More 'nearly' activity followed. Alen took seventh on the Finnish, after digging himself out of a snow drift, and on the Swedish Ingvar Carlsson and Claes Billstam were fifth, and Alen/Kivimaki sixth. Bernard Darniche was leading the Moroccan until he flooded the car in a river, while Alen and Waldegaard took a drier route to retirement by striking boulders out in the desert.

ESSENTIAL FIAT 124 SPIDER & COUPÉ

Two of the finest – and dustiest – moments in the 1974 campaign. Pinto/Bernacchini (above) headed the outstanding 1-2-3 finish on the TAP Rally in Portugal, and later in the season Bisulli/Rosetti (left) were the only crew to survive the Fiat team's decimation on the San Remo, finishing second.

Late-season dramas in 1974. Alcide Paganelli (above) runs wide at Bramham Park on the RAC, while on the Tour de Corse sixth-placed Fulvio Bacchelli (right) avoided the accidents that befell his team-mates; on this predominantly tarmac-surfaced event, a wrong choice of slick tyre compound meant that all the other works Fiat entries crashed.

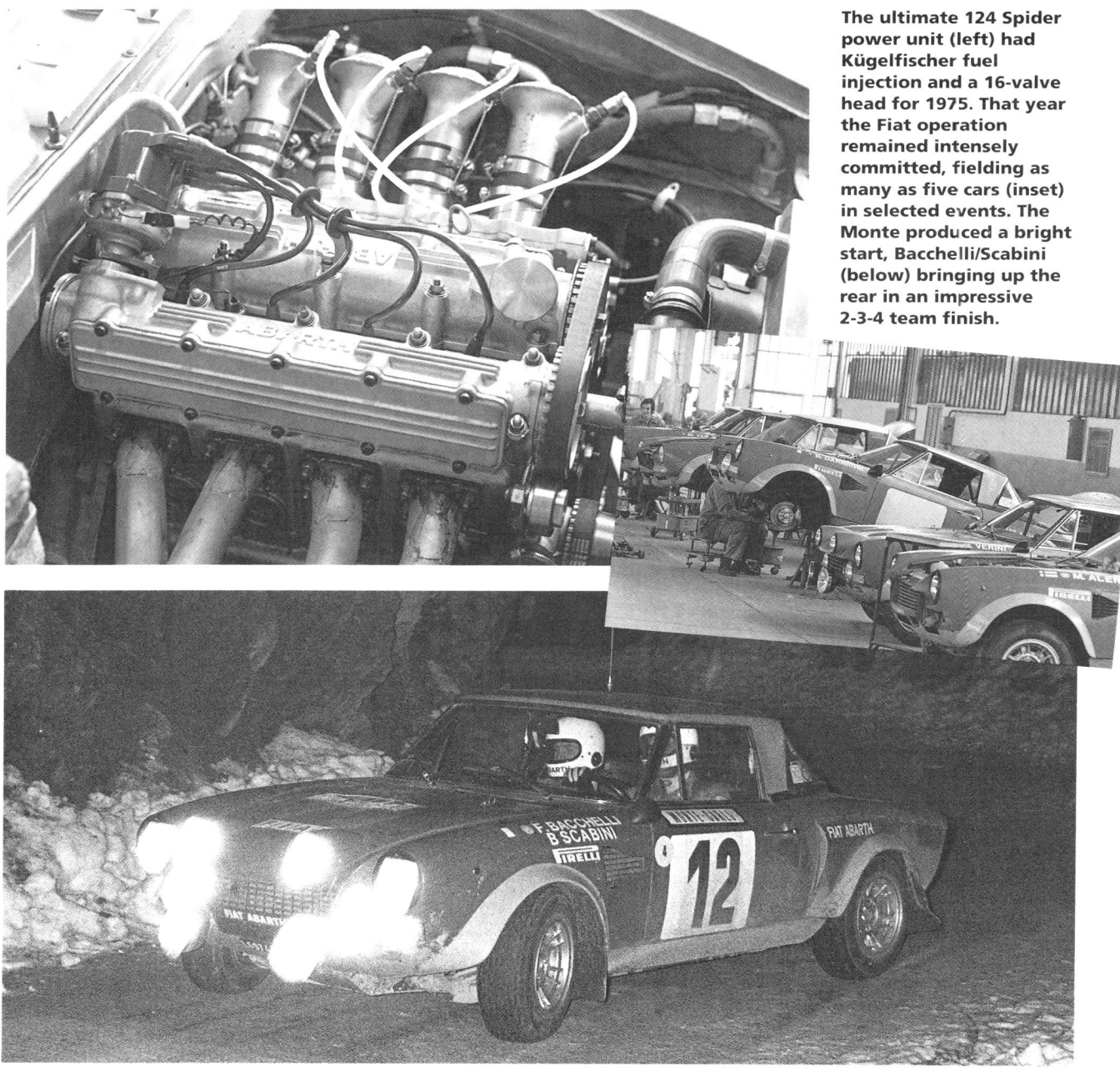

The ultimate 124 Spider power unit (left) had Kügelfischer fuel injection and a 16-valve head for 1975. That year the Fiat operation remained intensely committed, fielding as many as five cars (inset) in selected events. The Monte produced a bright start, Bacchelli/Scabini (below) bringing up the rear in an impressive 2-3-4 team finish.

Then came the first real success of the year when Alen (the 'New Finn') won the Portuguese TAP – it was his first World Rally Championship win. Even better, Mikkola (the 'Old Finn') was second in the oldest works car, which had seen such a hard life it let in dust – a real problem for driver vision on this event – through cracks in the floorpan. The tussle between the two men was fast and furious, with Alen pulling out a clear lead on the loose stages, only for Mikkola to close, then finally blow his chances by spinning.

In a feature published by *Classic and Sportscar* magazine in January 1985, Alen recalled his Spider days with affection, if in somewhat stilted English: 'I was a young man and it [the car] suited my aggressive driving style: very sideways all the time. It was very much a racing car type of design with very short shocks. The car was very

The 124 Spider's last World Rally Championship victory came in 1975 on traditionally fruitful territory in Portugal, where upcoming star Markku Alen achieved his first top-level win. These shots show the tyre variety used on mixed-surface events, although it is curious to see Alen's car wearing knobbly rubber for tarmac pirouettes at Estoril race circuit and grooved slicks for a dusty stage.

easy to drive, unlike today's prototype Lancia. I enjoyed the car tremendously.'

By the time the season moved to San Remo, Fiat was still in contention to win the World Championship if the Lancia team failed. Four cars were entered, all equipped with 16-valve heads. In addition two had the fuel injection which made the Spider the equal of the Stratos on some stages. Yet there were too many problems. Alen's rear suspension failed just after the start, Paganelli crashed, Cambiaghi's engine blew and only Verini/Rosetti survived to come in second behind the Waldegaard/ Thorzelius Stratos, with the Alpine of Therier third.

After that it was really all over. No Spiders were entered for the Tour de Corse and Lancia took the World Rally Championship with a win by Darniche/Mahe in a Stratos, while at the British RAC the Spider of

Climate contrasts in 1975: Ingvar Carlsson and Claes Billstam finished fifth in Sweden, but Bjorn Waldegaard and Claes-Goran Andersson had to retire in Morocco.

Verini/Rosetti was pushed down to eighth by the dominant Ford Escort RS 1600s and 1800s.

Once again there were consolation awards for the Fiat publicity machine to trumpet. Spiders took first, second and third in the European Drivers' Championship, with Verini leading Bacchelli and Jaroszewicz, while Tacchini/Simoni won the Mitropa Cup and Cambiaghi/Sanfront the Italian Rally Championship.

But by 1976 the Spider's competition days were numbered. Fiat was now gearing up for the debut of the 131 Abarth. Overall company management therefore decreed that Lancia should aim for the World Championship honours, while the ageing Spider lowered its sights to try to retain the European Drivers' Championship. One was entered in the Monte, but when the first 131 Abarth appeared in Morocco the Spider began being effectively eased out of the works picture.

The 131 Abarth, in turn, went on to be a resounding success, using 1976 to bed down before going on to win the World Rally Championship in 1977, 1978 and 1980.

Today Spiders with a competition history, either works or semi-works though outfits like the Jolly Club, are few and far between and draw big money from collectors.

THE LOST CLASSIC OF TURIN

The 124 Coupé, introduced in 1967 at the Geneva Show, was in some ways even more impressive than the Spider – and certainly much more versatile. Yet, 30 years on, it has almost disappeared from the roads, mostly claimed by the dreaded rust bug that plagued Fiat's products at the time. Strangely, as the open car was never sold here, in Britain in the 1990s you are much more likely to see a Spider than a Coupé, even though the latter sold here in large numbers.

If there was ever a forgotten classic, this is it. In its prime it was the finest family man's four-seater coupé you could buy and inspired a raft of imitators, Ford's Capri and Opel's Manta coming as its closest rivals in the mass market. With four real seats and a decent boot, it was a true grand tourer, although Fiat never demeaned it with the GT appellation that had already been debased by inferior machinery.

The company also did not overcharge for such excellence. The original AC series, at £1391 after tax, undercut all rivals bar the MGB GT, which was

Launched a year after the Spider, in 1967, the 124 Coupé was received enthusiastically by press and public. Superb styling, roomy interior, outstanding handling and decent performance – it was hard to fault.

considered far less civilised and sophisticated. Perhaps the only car approaching it was the BMW 1600, a quality compact with verve and ability, yet let down by its matronly appearance. In comparison the Reliant Scimitar looked an amateur job, as did the nimble and much faster Lotus Elan SE Coupé, while the Volvo P1800 looked and felt old, even if its 2-litre engine gave a healthy turn of speed and it was driven by Roger Moore as The Saint.

The standard model also spawned a couple of low volume 'in-house' imitations, Moretti building a coupé inspired by the Dino, but based on the 125 saloon, while Vignale offered the Samantha, also 125-based. About 30 of these cars were imported into the UK.

The Coupé had the same 95in wheelbase as the 124 saloon, rather than the slightly truncated platform of the Spider, with a boot that extended under the rear window parcel shelf, while the spare wheel was tucked away in a well. Apart from having four seats, it was quieter and more refined than the Spider, yet retained all the outstanding handling finesse and most of the performance of the open car.

While it might not have been so obviously appealing, the Coupé's crisp styling, low belt line and slender pillars gave it an unpretentious elegance not unlike its contemporary, the Lancia Fulvia. Yet, to their shame, the Fiat designers gradually ruined later models through a series of changes that supposedly kept the shape 'modern'.

Pininfarina had tried to tempt Fiat with a fixed-head version of the Spider at Turin in 1966. But the 124 Coupé stayed in-house, just like the smaller 850 Coupé, while the 850 Spyder was styled and built by Bertone. The 850 was confusingly sold in different markets as 'Spyder' and 'Spider', which by now was replacing the original spelling for foreign markets, and by the time of the 124 had become standard. The word 'Spyder' was coined at the beginning of the century to describe a light two-seater car, sometimes with a precarious looking third seat behind. It had been revived by some Italian manufacturers in the 1950s.

At Bertone, Fiat gave the job to the father and son team, Mario and Paolo Boano, who had controlled Fiat Centro Stile at Via La Manta since 1958. Mario, the father, had an impeccable pedigree. He had worked pre-war for Pinin Farina, then become part-owner of Ghia, where he had designed and built wooden styling bucks for other coachbuilders. In 1954 he had branched out under his own name, building among other projects a pair of Ferrari 410 Superamericas in 1957 and a short series of 250 Coupés to a Pininfarina design between 1956-58. At that point he had kept the business in the family by

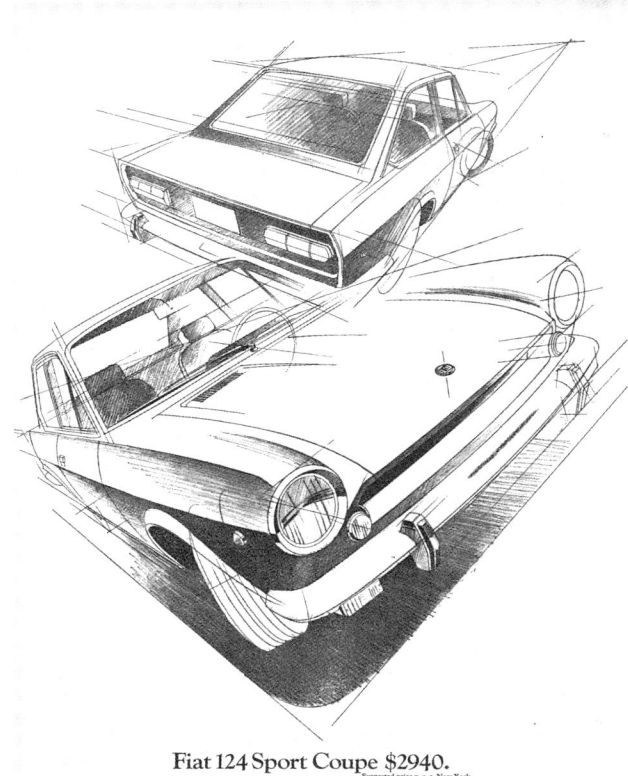

North American advertising: fine sketches really do justice to the Coupé, emphasising its exceptional style.

handing it over to his son-in-law and, along with his son, joined Fiat.

The pair's design for the 124 Coupé attracted a rapturous reception from the motoring press. Technical sophistication, classic looks, affordability – was this have your cake and eat time? You bet, they replied without hesitation. Particularly enamoured of the sensational handling, road testers and pundits were calling the Coupé a classic in print from the beginning, delving deep into their vocabularies in search of suitable superlatives.

'It is very hard to find out why the Coupé should have such out-of-this-world roadholding, but the fact remains that, so far as cornering goes, it has little to touch it,' the authoritative *Autocar* effused.

Technically, as the scribes then went on to point out, the Coupé differed very little from the Spider, launched the previous year. It had the same handsome little double overhead cam in-line four, with its modern cogged belt cam drive and twin-choke Weber 34DFH4 carburettor. Compression ratio was 8.9:1, for a net output of 90bhp,

North American photography: breathtaking winter backdrop for a Canadian-registered car at Niagara.

or 96bhp as quoted in North America, with other highlights of full-flow and by-pass oil filters and a thermostatically controlled electric fan. Impressively, when US imports began in 1968 the engine met all emission requirements in standard form.

The all-round 9in disc brakes were the same as the Spider's, with the Fiat servo providing light, but progressive, retardation and total indifference to hard and repeated use. Even the handbrake worked well, a rarity on early all-disc systems. Tyres again were identical – 165 SR 13 radials, either Michelin or Pirelli, on four-stud, 5in wide steel rims. The suspension was also the same – unequal length wishbones and coil springs at the front and live beam axle at the rear, located by trailing arms, coil springs and a Panhard rod. Weight distribution was 54%/56% front/rear, with the slight extra weight meaning the Coupé rode better than the sometimes choppy saloon and certainly equalled the Spider, which in turn rode brilliantly for an open sports car.

But above all it was the handling that was so uncannily good. 'It seemed to make little difference if the roads were wet or dry, we never reached the limit of grip, even at MIRA,' *Autocar* reported in awe. 'Going around the road circuit faster and faster became ridiculous, for even when we deliberately yanked the wheel at the apex, long after the car was set up and drifting hard, it simply turned in tighter...on public roads it seems no corner can be taken too fast and, for want of a better cliché, it really does feel as if the car is running on rails.'

Motor was a little more reserved, criticising the 'Alpine' gear ratios and complaining of a sometimes harsh ride. Yet its summing up was favourable: '...the 124 Coupé impressed us as an appealing car and an important newcomer in a domestically neglected market with much to offer the enthusiast wanting a reasonably priced sporting car with four seats.'

Autocar's tests gave a best top speed in fourth of 105mph, just a few revs short of the power peak, and 0-60mph in 12sec. Bigger-engined rivals like the Reliant Scimitar and Volvo P1800S had the edge in a straight line, the magazine concluded, but the 124 could easily out-corner them, thanks to its tenacious roadholding and four-square poise.

The main deficiency was the 1438cc engine's weak torque of 80lb ft (net) at 4000rpm, which became more obvious in the heavier Coupé when the rear seats were occupied. To compensate, Fiat closed up first, second and third, and initially offered the car only as a buzzy four-speeder that pulled 16.1mph per 1000rpm in top. This made it nifty off the mark, but meant it tended to go through an annoying exhaust boom at 70mph. From chassis number 005752 five speeds were then offered as a £38 option, yet only with a 0.913 reduction in ratio, which dropped the revs a mere 500rpm. Still, the Coupé's

Good handling was just one of the Coupé's many virtues. 'It is very hard to find out why the Coupé should have such out-of-this-world roadholding, but the fact remains that, so far as cornering goes, it has little to touch it,' stated *Autocar* magazine.

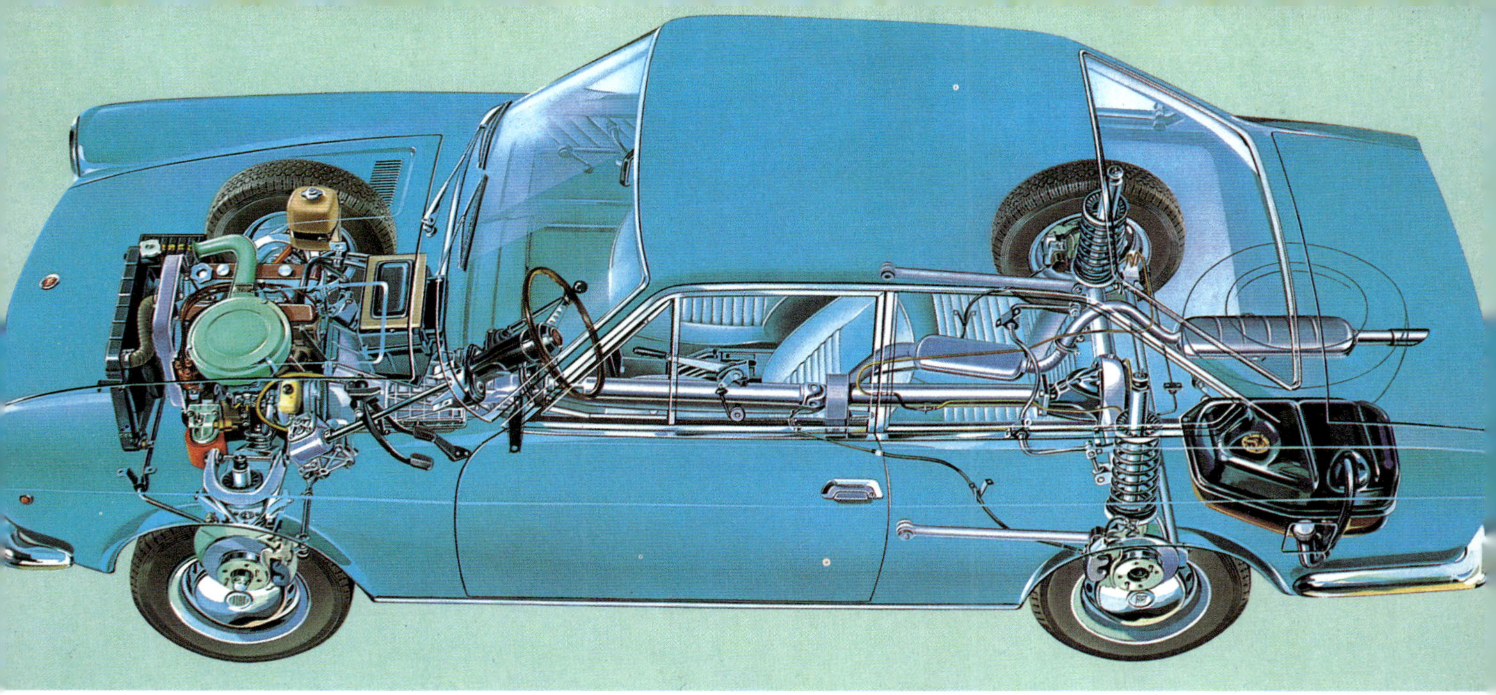

Although Coupé was mechanically identical to the Spider, interior was very different: four seats and a special dashboard – but trim is still shiny black vinyl on this remarkable 1438cc AC model from 1969.

65

Mild revisions in November 1969 – after little more than two years in production – brought the BC model, with quad-headlamp frontal styling reminiscent of the upmarket Dino. Optional alloy wheels are seen.

fifth gear was usable at around 35mph, while the rev counter was yellow-lined at 6600rpm to give maxima of 28, 50, 78, 106 and 100mph, showing that fifth was effectively an overdrive.

Intriguingly, for the Italian market, Fiat offered early Coupés with a form of selective automatic transmission, awkwardly named 'Idroconvert' or 'Idromatic'. This comprised a three-speed gearbox, with hydraulic torque converter and hydraulically controlled clutch, activated by an electro-magnetic switch on the gear lever. The system, built by Ferodo, had first been seen on the 850 saloon, where it had proved both expensive and unpopular.

Apart from this unfortunate loop, few rivals could equal the Fiat's level of equipment at the price – coat hooks, a map light over the parcel shelf, cigarette lighter, anti-dazzle mirror, twin roof lamps for rear seat passengers and twin air horns. There are few modern cars with a rheostat control for the wiper speeds, a system pioneered on the Rover 2000, although the crude manual plunger for the screen washers struck a jarring note. Although these accessories sound mundane today, in the mid to late 1960s ordinary cars were only just getting heaters and windscreen washers as standard equipment. Meanwhile, and most opulent of all, you could specify a handsome set of Cromodora alloy wheels, specially cast for the car and costing an additional £61 9s 2d.

Inside, the wood-rim steering wheel added to the impression of a cut-price Ferrari. Stalks on the column controlled lights, indicators and wipers, while the speedo and rev counter were set in a restrained wood-panelled binnacle, along with Veglia Borletti gauges for oil pressure, water temperature and fuel level. The heater was controlled by levers between the seats, with a rocker switch for the fan behind the centre console tray.

The light and airy interior of an early 124 Coupé has today an ambience of 1960s chintz with its acres of period black vinyl. By coupé standards there is excellent headroom, while the firm, short-backed front seats are more comfortable than they look, embracing both hips and back well. The rear seats, too, are shapely and supportive, with sufficient knee room for adults for medium length journeys.

Top of the hit parade

When the Coupé arrived in Britain in 1968, at the height of the Swinging Sixties, Fiat still had an image problem. Dealers were sparse and the motor trade regarded the company, which imported direct rather than through a separate importer, as a second-string franchise, rather like Eastern bloc imports are regarded today. The little 500 and 600 models tended to sell to working class people trading up from motorcycle combinations, while at the other end of the scale the big 2300 saloon and coupé dribbled out in tiny numbers to motorists wanting something a bit different and not afraid to pay. Most left the Jack Barclay showrooms in Mayfair for a life among the country house set.

While the 850 and 124 saloons gave Fiat a more mainstream appeal, the Coupé brought in a completely new type of buyer – the 1960s equivalent of a 1980s 'yuppie', interested in the look of the car and impressed by the reviews in the motoring press. Suddenly Fiat had such an 'in' car, and there was such a shortage that for a

Interior details changed on the BC: better seats with cloth panels, the addition of a clock on the black instrument panel, and eyeball air vents on the centre console.

long time second-hand examples commanded a considerable premium over new – a novel, and perhaps unique, accolade for the company.

In November 1969 Fiat introduced the BC series with 'improved' styling – although few pundits visiting the Fiat stand at Turin were impressed. Four headlights had now appeared, giving the front end a slight Dino look, while the simple AC grille had grown to full width and the new bonnet had been reduced to a single vent. The round indicators, previously next to the grille, were replaced by rectangular ones niched into the lower part of the bumper, while the Fiat badge moved from the bonnet to the centre of the grille. At the rear a revised light cluster replaced the units the AC had previously shared with the Lamborghini Espada, but otherwise there was little to tell the new car apart from behind. Overall it was now perhaps less distinctive, but still restrained and well-judged, without excessive brightwork.

Inside, the seats were more contoured and sported central cloth inserts, while the unchanged steering wheel now adjusted for height, helping taller drivers to achieve a less 'Italian ape' driving position. Yet the seat could still have done with more rearward adjustment, as the pedals were still too close for some even when the steering wheel was at full stretch. The facia now contained a clock, but was otherwise the same, while the heating system had been improved. Nonetheless, the fan was still noisy and critics deemed the ventilation inadequate – a common complaint about Italian cars.

Under the bonnet there was a new 1608cc twin-cam, with a longer stroke, as fitted in the 125 saloon and using twin Weber IDF carburettors and reprofiled cams to raise power output to 110bhp at 6500rpm. In service this twin-carburettor arrangement was to prove notoriously difficult to set up if the induction for each cylinder was allowed to get out of 'sync' – a malady usually indicated by poor running and excessive thirst. But when it was running right the reward was even crisper throttle response and eagerness to go to the red line.

Five speeds were now standard, but by lowering the axle ratio to 4.3:1 against the 4.1:1 of the still-optional 1400cc model, Fiat created a curious set of ratios. First, second and third were too low and too far apart, giving aggressive sprint gearing. First ran out at a mere 27mph, second at 48mph and third at 73mph, while fourth took the car to 101mph. But even though the gap between it and fifth was usefully wider than on the 1400, it was still hardly a restful overdrive ratio with only 17.5mph per 1000rpm. The payback, however, was the smooth engine, which enabled the car to cruise at speed without sounding strained.

It was at the upper end that owners felt the real

BC differences are clear from the front, while the rear is distinguished by larger light clusters and a telling badge – under the bonnet was the new 1608cc twin-cam.

benefit. The 70-90mph time was cut in half to 10.6sec, and while the old 1400 had been unable to pull a time to 100mph that was worth recording, the 1600 reached the magic figure in a respectable 43sec.

Critics did sense some sharpness had been sacrificed in the handling, thanks to the replacement of the torque tube by a conventional prop shaft, along with a pair of shorter radius rods with bigger bushes. The change was, however, entirely necessary. As with the Spider, rear axles had kept breaking due to misaligned radius rods,

THE LOST CLASSIC OF TURIN

To most eyes, Fiat mucked up the Coupé's styling for the CC model, introduced in August 1972.

particularly on rough roads. Road testers said the new set-up introduced more roll and understeer, but it would be wrong to overstate the case. The 124 was still superbly safe and responsive, with lots of cornering power in reserve in the dry, and wonderful balance in the wet. Flicked into a greasy roundabout, the understeer could easily be overcome to produce a beautifully controllable slide, while poor surfaces still hardly affected the well-located rear suspension.

'To flick the car quickly along a winding country road is to enjoy the 124 Sport at its very best,' as *Motor* explained.

By way of comparison, the magazine tested the 1400cc version, which still sold for £100 less. Interestingly, it proved quicker than the original AC 1400 tested three years previously. It was also hardly any slower than the 1600, raising the suspicion that the bigger-engined test car had been down on power. Top speed was very little behind, at 104mph, and it was only in fourth gear that the 1600 really scored.

The 9in disc brakes, with Fiat-Bendix servo, were less well received. Although apparently much as before, *Motor* criticised them for sponginess and over-sensitivity that hindered heel-and-toe changes, along with lack of progression, particularly on release. However, this may just have been a peculiarity of the magazine's test car. *Autocar*'s example, registered DYT 719J, suffered no such problems, the brakes remaining light to operate and nicely progressive.

With a top speed of 109mph, the 1600 was one of the quickest coupés in its class. But, carrying the burden of pre-Common Market import duty, it now faced stiffer opposition. Ford had introduced its beefy 3-litre Capri, which sold for less (although it was not regarded as such a rewarding driver's car), the 2-litre BMW 2002 had become a benchmark for rapid saloons, and there was further competition from also-rans like the Sunbeam Rapier H120, which the less enlightened might have viewed as better value for money. But none of these really worried Fiat, and the Coupé continued to sell strongly all over Europe, usually with a waiting list.

In North America, however, it was always rather overshadowed by the Spider, selling steadily rather than strongly against formidable opposition from Datsun's 240Z, which was much faster, as well as having six

Practical improvements for the CC Coupé. Interior now had aluminium-finished facia with speedo and rev counter recessed to prevent glare, while boot access was transformed.

cylinders. In the States the performance of the detoxed engine was down, 0-60mph now taking 12.4sec against the 11.4 of the original 1968 car. Fuel consumption was also higher, 24mpg against 22 according to *Road & Track*, while the price had gone up by $300. Yet the Coupé was still something of a bargain, undercutting all its rivals in the medium-sized bracket, bar the Opel GT. In 1971 *Road & Track* compared it with the new Datsun 240Z, the Triumph GT6, the MGB GT and the Opel, ranking it a close second behind the Japanese car and giving it top marks for handling and practicality. Meanwhile, the magazine reminded its readers, the 124 was the only real four-seater in the group.

Style on the slide

It was the CC Coupé, announced in August 1972 for the 1973 model year, that ushered in radical styling changes as Fiat attempted to boost the appeal of a now six-year-old design. At the front the bumper became split-level, the two outer sections incorporating enlarged indicators, while a slimmer, centre section sat between two rubber-faced overriders. The twin headlights now occupied their own panels on either side of the rectangular grille, which was enclosed inside a slightly bizarre chrome surround, while the bonnet had twin swage lines from front to back, along with a larger air intake.

The new triangle of black plastic emphasising the 'C' pillar was actually a slatted grille for through-flow ventilation. At the back the rear light clusters were enlarged and restyled, while the boot lid rim was lowered to the level of the rear bumper to make access to the luggage compartment easier. This stifled the moans of middle-aged golfers, who had long complained about having to heave their clubs over the previous high lip. New side mouldings in rubber protected the car against parking knocks, while Fiat claimed the black 'acrylic finish' on the sills would protect this vulnerable area from the tendency to corrosion that was actually the car's main problem. Wheels and tyres were the same as before, but bedecked with fancier hub caps.

Inside, the instruments were now deeply hooded to avoid glare and sat in an aluminium-finished facia. The

Coupé was a huge success in Britain. At first there was such a long waiting list that second-hand examples commanded a price premium – very unusual, and unprecedented for a Fiat.

With its bluff nose, the CC Coupé appeared less aerodynamic, but wind tunnel work at Pininfarina proved otherwise.

glove box had gone in favour of a moulded trough with a parcel tray below, while the updated electric screen wash now worked off a column stalk. Meanwhile, the hand throttle – a quaint feature of better Italian cars – remained useful for keeping the engine running on cold mornings. Garish black and white checked trim, in nylon, reflected the tastes of the time and destroyed the air of restraint which had once made the interior so pleasant.

The mechanical changes were due to rationalisation with the new 132 saloon rather than a quest for more performance. By using the same new block and head castings, with wider cylinder spacing, the existing 1592cc unit now had the option of a revised 1756cc twin-cam, bored out from 80 to 84mm and with the stroke reduced slightly from 80 to 79.2mm. It was also possible to specify a four-speed gearbox, which was actually the old one with fifth removed.

Both models breathed through a single twin-choke Weber carburettor, with power climbing a modest 8bhp on the bigger-engined version to 118bhp. But as weight and gearing remained the same, performance was hardly affected. If anything, the 1800 was both thirstier and a shade slower than the old twin-carb 1600, particularly in the upper range of acceleration in fourth and fifth.

Even so, the car still squared up well against its Euro Coupé rivals, with the same excellent handling and strong, sweetly delivered performance from the superb engine. The only fault was with the five-speed 'box, shared with the 132 saloon. Although this was all-new, it still had the closely-stacked ratios that gave the car such an urgent feel at low speeds.

In refinement only Audi's fine 100S Coupé was now nudging ahead, but that was more pricey and not such good fun to drive on twisty roads, where the rear-wheel drive Fiat still excelled. With the rear anti-roll bar now reinstated, and the brakes modified, some of the original model's handling prowess had been regained, a consolation for the way the styling was now criticised for being ruined by such unseemly fiddling.

At a shade over £2000 in Britain, it was still an attractive prospect as a family man's sports car, although many were bought by well-heeled individuals, often women, who were looking for a stylish second car rather than an everyday hack.

By the time the 124 Coupé passed quietly away in 1975 a total of 279,000 had been built over eight years – about 80,000 more than the Spider that was to live on for a further decade. Discounting the 128 Coupé and 3P, which were really nothing more than sporty hatchback versions of the front-driven 128, there were to be no more coupés from Fiat for the next two decades. As the Lancia Beta Coupé had now come on stream after Fiat had bought Lancia (for a nominal 1 lira per share) in 1969, the management deemed an in-house competitor superfluous, a decision that was probably correct. The Beta, although not as outstanding as the 124 had once been, was still one of the best of the new type of modern front-wheel drives.

In an interesting postscript, 124 production continued until 1977 in Spain, where the car had been manufactured as the Seat 124 Coupé since 1974. Marketed initially with the 105bhp 1756cc 132 engine, it was then sold with the high-compression, high-lift twin-cam.

It was not until 1993, after the quiet coupé market of the 1980s, that Fiat re-entered the field with a coupé designed in the Fiat styling centre by the American, Chris Bangle, but built at Pininfarina. Based on the Tipo and costing £80 million to develop, it was bold and radical, and although its individualistic styling met a mixed reception, the critics deemed its performance (particularly in turbo form) and impeccable front-wheel drive manners as stunningly good. And for British enthusiasts it was imported with right-hand drive, unlike the Barchetta.

Two decades apart: modern Fiat Coupé with the 'lost classic'. Opinions differ about which would win a beauty contest...

Spiders and Coupés Today

Pininfarina Spidereuropa in a Spider graveyard. The scene is at DTR, Britain's leading specialist for these cars: wrecks are there to be pillaged for elusive parts.

Many chauvinistic British enthusiasts dismiss the 124 Spider as an Italian MGB, thereby underplaying the ability of this great all-rounder. Yet there is a grain of truth in the oft-quoted cliché that for every hardcore British fanatic regarding it as a troublesome baby Ferrari, a dozen European and North American fans see it as a cheap and unpretentious fun car.

Which model is the one for you depends on your purpose and finances. For the connoisseur, the *Stradale* Abarth is still the ultimate. But there are many imitations and identity must always be checked thoroughly. Genuine Abarths come up for sale reasonably regularly, but their rarity puts them into a lofty price bracket.

For a more practical – as well cheaper – fun car, the most desirable and collectable are the pretty chrome-bumper models, made up to 1974. The purist goes for an early AC (with torque tube), which, although it has engine torque in short supply, is very sweet mechanically and has a fine 1960s interior. These early cars, equipped with an anti-roll bar, also handle better. The most responsive performance comes from the 1608cc models with twin carbs, but these are difficult for the private owner to keep in tune, tending to suffer from both thirst and annoying flat spots.

However, comparatively few chrome-bumper AS, BS and CS versions have survived and they are now cherished collectors' items, meaning you are much more likely to end up with one of the big-bumper models. For everyday use the injection examples are torquey, untemperamental and good motorway cruisers, with the late Spidereuropas feeling particularly 'modern' to drive. The 2-litre carburettor models, only made for one year, are perhaps the least pleasant in standard form, but respond to tuning better than the 1600s.

Meanwhile the supercharged VXs, unheard of outside mainland Europe, are expensive if they do become available. Most have ended up in Germany, where the Spider has a highly enthusiastic following. The American-converted 2-litre turbos have good mid-range acceleration, but the turbo tends to put excessive strain on the bottom end of the engine and the rest of the drive train, knocking out both bearings and synchromesh – and turbocharger parts are difficult to obtain.

In Britain a small network of specialists now caters for the 1500 or so cars that have been personally imported. Importation, usually from California, is still an option for the private buyer, but fraught with difficulties. 'Genuine' Californian Spiders, the mythical rust-free cars, are not always what they seem. Many are sourced from out of state, while cars from the easiest route of Florida can be as rotten as those in the UK, thanks to the humid climate.

It should also be remembered that cars from North America come loaded with power-sapping emission equipment. You either have to live with it or, expensively, bring back European tune by fitting high-compression pistons and a four-branch manifold. Another option is to fit a Coupé engine, but these cars are now so thin on the ground they are hard to find.

Many novices attempting an import fall foul of unscrupulous sellers and end up saddled with a 'dog', often because of badly-repaired accident damage. There is no MoT in the States, while the Spider is a cheap car, usually owned by younger drivers who sort out problems as cheaply as possible. Furthermore, there is always the danger of becoming ensnared in customs' red tape that can leave you several hundred pounds out of pocket. Far better to buy a Spider which has already been sorted on this side of the Atlantic.

If left-hand drive is not acceptable, conversion to right-hand drive is possible as the dashboard is handily symmetrical, with the glove box almost perfectly mirroring the instrument console. The steering column, idler and pedal box can be taken from the Coupé.

When inspecting a Spider look for buckled chassis legs in particular. The cross members, where they bolt to the chassis legs on two mounting points, are also an important safety check as the bolts tend to work loose under braking and cornering. Eventually a hairline crack can develop and the cross member break away, with predictably sad and expensive results. Look out also for worn ball joints on the front suspension, as correct greasing is often overlooked.

And if you do not like the handling of your big-bumper Spider, you can always lower it and fit gas-pressure shock absorbers to regain much of the poise of earlier cars. There is also plenty of room to fit modern low-profile tyres that will give you unshakeable grip.

Engines can suffer from rumbly bottom ends, and on the 2-litre cars the unit is so low-slung that sump damage is common, affecting the oil pick-up and meaning that oil pressure suffers, although the engine still runs. Having said that, if the engine seems healthy enough you should not always pay too much attention to the gauge, as the sender unit frequently goes wrong, relaying misleading and alarming messages.

Owners also tend to abuse these durable engines by letting them run low on oil or water, while leaks can occur from the lower cam gasket on the exhaust side. On the 2-litre engine the tensioner bearing on the cam belt can seize, indicated by a whirring that is sometimes difficult to pick out from the general noise. Failure to do so will result in the cam belt itself snapping, thereby forcibly drawing the defect to the attention of even the most dozy owner.

Gearboxes are strong on earlier cars, but weaker on later models which sourced them from Seat in Spain. Hubs and synchros are expensive to replace but need to be tackled if you are doing a proper rebuild and want the gearbox to last. It is a common trait for an owner to uprate the engine but ignore the 'box, which may have a huge mileage. Front oil seal failure causes fluid loss and premature failure, often signposted by jumping out of fourth gear or losing it altogether.

US Spiders tend to have been abused in this department because many American drivers are unsympathetic to manual changing, but the gearboxes on British Coupés are usually fine. Automatic Spiders are almost unsaleable outside America: DTR, the British specialist in Twickenham, Middlesex, have to convert imported specimens to manual to make them saleable.

A useful tip when driving any car is to sense for a lack of feel in a straight line, which may indicate wear in the (expensive) steering box. Clutch problems are also common. The clutch becomes heavy, fracturing the pedal, at which the cable bends the fork and then snaps. The propshaft, too, is a source of trouble. In first and second gear vibration can occur from the centre support bearing, which rarely gets changed. This movement in the propshaft then affects the differential, which eventually fails. On the later 2-litre cars the diff is particularly

Spider rallying exploits are still recalled in historic events. This is Mike Wood's ex-Jolly Club Abarth, to Group 4 specification.

Spider appeal is truly international. This club line-up, fronted by a rare and highly desirable 2000 Turbo, is in Holland.

expensive to rebuild properly as it is assembled in the casing, rather than being a separate unit.

Seized brake calipers can cause the car to pull to one side, which is cheap to cure. But look out that this is not being caused by a damaged wishbone or cross member. And when the caliper body distorts, as it can with age, and causes the brakes to grab on, do not make the mistake of many owners who rebuild them, only to experience the same problems a few weeks later. The only answer is new or exchange calipers.

Above every other consideration, though, is rust. Year for year, older cars are not so prone, as they were better built. But often there is no escape and if you see one area needing doing, you can be almost certain the rest will be as bad, no matter how tidy the car may appear.

Specifically, the rear arches are a massive water trap. Likewise the bottom section where the front wing goes up and over the wheel, with the corrosion usually extending into the door post and jacking point. Doors themselves are also prone to rot and it is worth knowing that only late-type 2-litre doors, with flush Ferrari 400 exterior handles, are still available. New rear wings cannot be obtained at any price, but pattern front wings are manufactured in the US and there is a limited supply of Pininfarina originals.

Outer sills, just covers that screw on, can be replaced cheaply, but if they do contain any holes it is almost certain the inners will be affected as well. And with panel prices on the high side, most owners will not have done a proper repair, ignoring, for example, the rear inner arch. Many just rivet in a section, allowing the rot to eat away further, unseen, until it reaches the point where you almost have to scrap the car and start again.

Beware of any Spider with missing exterior trim details. Items like headlamp bezels and bumpers are extremely rare and expensive, particularly for early cars,

Rebuild at DTR of a highly desirable power unit: 1608cc engine with twin Webers is from a 1972 Spider, with high-compression pistons and new camshafts contributing to power hike from 110bhp to 125bhp

with rear lights for ASs like the proverbial gold dust.

On American cars the fierce Californian sun has usually ravaged the interior, badly affecting the brittle plastic dash top, which is prone to cracking. Soft-tops need to be carefully checked for damaged or seized frames and tears in the fabric. Replacements are highly expensive and not always top quality. Missing interior parts can be difficult to find, although retrimming the vinyl is straightforward enough and second-hand parts can be bought relatively cheaply in the States.

In general, though, spares availability is good, although the Spider is still more of a challenge than its British sports car contemporaries. All the consumables for running a car from day to day are available off-the-shelf while, at the other extreme, even badges and trim for the rarefied *Stradale* can be sourced in Europe. A highly-regarded specialist in Germany is Holtmann & Niedergerke, who in 1990 cunningly bought up a line of ex-Fiat America stock that was languishing in Palm Springs. Their British agents are Middle Barton Garage, based in the Cotswolds village of the same name.

The pick of the Coupé bunch, to my mind, has to be the pretty first series AC. It has the sweetest looks, most sensational handling and great 1960s period presence. The biggest problem, though, is finding one. They are so close to extinction in the wetter European climates that one expert recently estimated less than ten were running in the UK. Those that have survived obviously tend to be cherished examples that have led sheltered lives.

Later Coupés have survived in greater numbers, thanks to Fiat belatedly realising the benefits of rudimentary rust-proofing and fitting some CC series cars with plastic wheel arch liners. Even in America the Coupé did not have as good a reputation for rust resistance as the Spider, adding credence to the rumour that the cars had been built of sub-standard Soviet steel. One way out, though, is a shopping trip to Spain where Seat made 9000 CCs under licence.

Mechanical bits should be no problem, but body panels are very difficult to find, with only reproduction wheel arch sections currently available. Anyhow, restoring a rotten Coupé does not really make sense. Values have been stagnant for years and sorting one out is a major undertaking. The cars suffer particularly around the scuttle

Three problem points. Twin-cam engine is generally robust, but occasionally on twin-carburettor units a valve can melt and drop into the cylinder if the front carb is running too lean. But rust is the main story: rampant inner sill rust is visible after outer panel has been removed; and poor repairs with filler are a common legacy on surviving cars today.

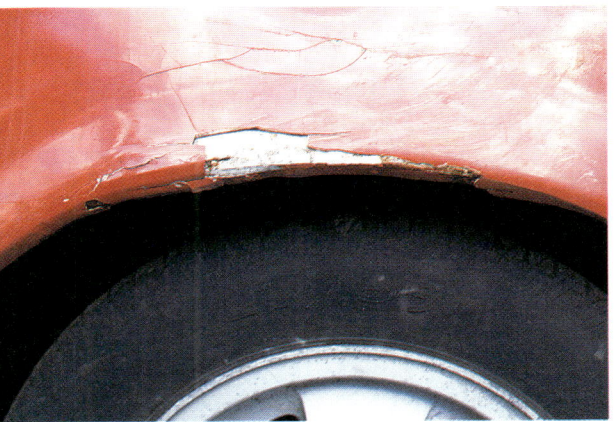

and screen pillars, a damp interior an immediate giveaway. Unseen rust holes under the window trim, along with poor screen-bonding, will be letting in water every time it rains.

Any repair to the scuttle area will mean removing both dashboard and screen, although the latter is at least still available. Rusty sills will seriously compromise the structural rigidity, while replacements now have to be fabricated from scratch. Check out the floorpan, the inner and outer wheel arches at both ends, the tops and bottoms of the door skins, and the valances under the bumpers. The area around the headlights quickly rotted out from new, especially on early cars without wheel arch liners.

However, many mechanical parts, although not usually common to all three models, are interchangeable, while a 2-litre engine conversion is perfectly sensible, especially if an early type is simply not obtainable. Attention needs to be paid to spring rates, as the 2-litre engine is heavier, while only the original carburettor gives enough bonnet clearance with the taller engine. No new interior pieces are available, but so many cars have been scrapped that second-hand parts are plentiful.

In many ways the 124 Spider and the Coupé, if you can find one, make ideal classic sports cars. Whatever you may hear muttered at an MG or Triumph meet, they are mechanically robust, giving you Italian dash and style with none of the heartache.

APPENDIX

Production figures

SPIDER

Spider AS	Aug 1966 to Jul 1970	32,327
Spider BS	Nov 1969 to Jul 1970	10,526
Spider BS1	Jul 1970 to Sep 1972	27,906
Spider CS	Sep 1972 to Jul 1973	10,730
Spider CSA	Sep 1972 to Sep 1974	1013
Spider CS1	Aug 1973 to Jul 1978	69,208
Spider CS2	Jul 1978 to Oct 1979	16,926
Spider CS0	May 1979 to Aug 1982	32,060
Spider CS0 Turbo	Apr 1981 to 1983	approx 700
Spider DS	Mar 1982 to Nov 1985	7450
Spider DS VX	Jul 1983 to Nov 1985	500

COUPÉ

AC	1967 to 1969	113,869
BC/BC1	1969 to 1972	99,500
CC/CC1	1972 to 1975	66,303

Technical specifications

SPIDER

SPIDER 'AS' (1966-69) Engine Four-cylinder in line **Crankshaft** Five main bearings **Bore x stroke** 80.0mm × 71.5mm (3.15in × 2.82in) **Capacity** 1438cc (87.5cu in) **Valves** Twin overhead camshafts **Compression ratio** 8.9:1 **Fuel system** Weber 34 DFH4 carburettor **Maximum power** 90bhp (net) at 6500rpm **Maximum torque** 80lb ft (net) at 4000rpm **Transmission** Four- or five-speed manual gearbox, all-synchromesh, 7.87in (200mm) diameter diaphragm spring clutch **Gear ratios** Fifth 0.91, fourth 1.00, third 1.49, second 2.30, first 3.75, reverse 3.65 **Final drive** 4.10:1 **Top gear speed per 1000rpm** 17.6mph (28.3kph) **Front suspension** Independent, double wishbones, coil springs, telescopic dampers, anti-roll bar **Rear suspension** Live axle, parallel trailing arms, Panhard rod, coil springs, telescopic dampers, anti-roll bar **Steering** Worm and roller, 15in (380mm) steering wheel, 34ft (10.4m) turning circle, 2.75 turns lock to lock **Brakes** Fiat-Bendix 8.95in (227mm) discs front and rear, Bonaldi vacuum servo **Wheels** Pressed steel disc, four studs, 5in rim **Tyres** Pirelli Cinturato 165 SR 13 **Length** 156.3in (3970mm) **Width** 63.5in (1613mm) **Height** 49.2in (1250mm) **Wheelbase** 89.9in (2283mm) **Ground clearance** 4.7in (119mm) **Front track** 53.0in (1346mm) **Rear track** 51.8in (1316mm) **Kerb weight** 2093lb (949kg) **Weight distribution** 56%/44% front/rear **Top speeds in gears** Fifth 106mph (171kph), fourth 104mph (167kph), third 77mph (124kph), second 50mph (80kph), first 30mph (48kph) **0-60mph** 11.9sec **Standing quarter-mile** 17.5sec

SPIDER 'BS1' (1969-72) As 'AS' except (but note 'BS' also available with 90bhp/1438cc engine): **Bore x stroke** 80.0mm × 80.0mm (3.15in × 3.15in) **Capacity** 1608cc (97.5cu in) **Compression ratio** 9.9:1 **Fuel system** Weber 40 IDF10/11 carburettor **Maximum power** 110bhp (net) at 6400rpm **Maximum torque** 101lb ft (net) at 3800rpm **Gear ratios** Fifth 0.91, fourth 1.00, third 1.41, second 2.18, first 3.80, reverse 3.53 **Top gear speed per 1000rpm** 17.7mph (28.4kph) **Kerb weight** 2190lb (993kg) **Weight distribution** 55.0%/45.0% front/rear **Top speeds in gears** Fifth 112mph (180kph), fourth 102mph (164kph), third 75mph (121kph), second 48mph (kph), first 28mph (77kph) **0-60mph** 12.2sec **Standing quarter-mile** 17.8sec

SPIDER ABARTH 'CSA' (1972-74) As 'BS' except: **Bore x stroke** 84.0mm × 79.2mm (3.31in × 3.12in) **Capacity** 1756cc (107.2cu in) **Compression ratio** 9.8:1 **Fuel system** Twin Weber 44 IDF dual-throat carburettors **Maximum power** 128bhp (DIN) at 6200rpm **Maximum torque** 117lb ft (DIN) at 5200rpm **Transmission** 7.92in (201mm) diameter diaphragm spring clutch **Gear ratios** Fifth 0.88, fourth 0.881, third 1.36, second 2.10, first 3.66, reverse 3.53 **Final drive** 4.3:1 **Top gear speed per 1000rpm** 17.6mph (28kph) **Rear suspension** Independent with lower links, radius rods, coil springs, telescopic dampers, anti-roll bar **Brakes** Fiat-Bendix 9.08in (230mm) discs front and rear, dual circuit, servo **Wheels** Magnesium alloy, four studs, 5.5in rim **Tyres** Pirelli Cinturato 185/70VR 13 **Length** 163.1in (4142mm) **Front track** 53.2in (1351mm) **Rear track** 52.0in (1320mm) **Kerb weight** 2067lb (938kg) **Weight distribution** 58.0%/42.0% front/rear **Top speed** 118mph (190kph) [manufacturer's claim]

SPIDER 'CS1' (1972-78) As 'BS' except (but note 'CS' also available with 105bhp/1592cc engine, Sep 1972 to Jul 1973): **Bore x stroke** 84.0mm × 79.2mm (3.31in × 3.12in) **Capacity** 1756cc (107.2cu in) **Compression ratio** 9.8:1 **Fuel system** Weber 34 DMS carburettor **Maximum power** 118bhp (DIN) at 6000rpm **Maximum torque** 111lb ft (DIN) at 4000rpm **Transmission** 7.92in (201mm) diameter diaphragm spring clutch **Gear ratios** Fifth 0.87, fourth 1.00, third 1.36, second 2.05, first 3.61, reverse 3.53 **Final drive** 4.3:1 **Top gear speed per 1000rpm** 17.6mph (28kph) **Brakes** Fiat-Bendix 9.08in (230mm) discs front and rear, dual circuit, servo **Length** 163.1in (4143mm) **Front track** 53.2in (1351mm) **Rear track** 52.0in (1320mm) **Kerb weight** 2365lb (1073kg) **Weight distribution** 53.0%/47.0% front/rear **Top speeds in gears** Fifth n/a, fourth 95mph (153kph), third 72mph (116kph), second 47mph (76kph), first 27mph (43kph) **0-60mph** 14.8sec **Standing quarter-mile** 20.0sec

SPIDER 2000 'CS2' (1978-79) As 'CS1' except: **Bore x stroke** 84.0mm × 90.0mm (3.31in × 3.54in) **Capacity** 1995cc (121.7cu in) **Compression ratio** 8.1:1 **Fuel system** Weber carburettor **Maximum power** 80bhp (net) at 5000rpm **Maximum torque** 104lb ft (net) at 3000rpm **Transmission** Optional automatic (GM three-speed) **Gear ratios** Manual: fifth 0.88, fourth 1.00, third 1.36, second 2.10, first 3.66, reverse 3.53. Automatic: third 1.0, second 1.48, first 2.40 **Final drive** 3.9:1 manual, 3.583 automatic **Top gear speed per 1000rpm** 19.4mph (31kph) automatic **Wheels** Alloys optional **Tyres** Pirelli P3 165 SR 13 **Top speeds in gears** Manual: fifth 102mph (164kph), fourth 102mph (164kph), third 79mph (127kph), second 52mph (84kph), first 30mph (48kph). Automatic: top 100mph (161kph) **0-60mph** 10.6sec manual, 12.2sec automatic **Standing quarter-mile** 18.1sec manual

SPIDER 2000 INJECTION 'CS0' (1979-82) As 'CS2' except: **Compression ratio** 8.2:1 **Fuel system** Bosch L Jetronic fuel injection **Maximum power** 102bhp (net) at 5500rpm **Maximum torque** 111lb ft (net) at 3300rpm **Steering** 3.5 turns lock-to-lock **Brakes** Fiat discs, 10.83in (275mm) front and 8.9in (225mm) rear **Wheels** Cromodora 14in alloys, 5.5in or 6in rims **Tyres** Pirelli P6 185/60 HR 14 **Kerb weight** 2337lb (1060kg) **Top speeds in gears** Fifth 109mph (175kph), fourth 96mph (154kph), third 72mph (116kph), second 47mph (76kph), first 27mph (43kph) **Standing quarter-mile** 17.9sec

SPIDER 2000 TURBO 'CS0' (1981-82) As Injection 'CS0' except:

APPENDIX

Compression ratio 8.1:1 **Fuel system** Bosch L Jetronic fuel injection with IHI Turbocharger **Maximum power** 120bhp (net) at 6000rpm **Maximum torque** 130lb ft (net) at 3600rpm **Top gear speed per 1000rpm** 17.1mph (27.5kph) **Kerb weight** 2385lb (1082kg) **Weight distribution** 52%/48% front/rear **Top speeds in gears** Fifth 104mph (167kph), fourth 90mph (145kph), third 69mph (111kph), second 44mph (71kph), first 26mph (42kph) **Standing quarter-mile** 17.1sec

PININFARINA SPIDEREUROPA/AZZURA 'DS' (1982–85) As Injection 'CS0' except: **Compression ratio** 8.2:1 **Maximum power** 102bhp (net) at 5500rpm **Maximum torque** 111lb ft (net) at 3300rpm **Final drive** 3.72:1 **Top gear speed per 1000rpm** 17.9mph (28.8kph) **Steering** Rack and pinion from 1985, 3.5 turns lock-to-lock **Wheels** Speedline 14in alloys, 6in rims; or Cromodora 13in alloys, 5.5in rims (Azzura) **Tyres** Pirelli P6 185/60 HR 14, or 165 SR 14 (Azzura) **Kerb weight** 2337lb (1060kg) **Weight distribution** 53%/47% front/rear **Top speeds in gears** Fifth 109mph (175kph), fourth 96mph (154kph), third 72mph (116kph), second 47mph (76kph), first 27mph (43kph) **Standing quarter-mile** 17.9sec

PININFARINA DS VX/SPIDEREUROPA VX (1983–85) As Injection 'CS0' except: **Compression ratio** 7.5:1 **Fuel system** Weber 36 DCA 7/250 downdraught twin-barrel carburettor and Roots type volumetric supercharger **Maximum power** 135bhp (net) at 5600rpm **Maximum torque** 152lb ft (net) at 3000rpm **Steering** Rack and pinion from 1985, 3.5 turns lock-to-lock **Wheels** Cromodora 14in alloys, 5.5in or 6in rims **Tyres** Pirelli P6 195/50VR 15 **Width** 64.0in **Front track** 54.1in **Top speeds in gears** Fifth 118mph (190kph), fourth 96mph (154kph), third 72mph (116kph), second 47mph (76kph), first 27mph (43kph) **0-62mph (100kph)** 8.6sec **Standing quarter-mile** 16sec

COUPÉ

SPORT COUPÉ 'AC' (1967–69) Engine Four cylinder in-line **Crankshaft** Five main bearings **Bore x stroke** 80.0mm × 71.5mm (3.15in × 2.82in) **Capacity** 1438cc (87.5cu in) **Valves** Twin overhead camshafts **Compression ratio** 8.9:1 **Fuel system** Weber 34 DFH4 two-stage carburettor **Maximum power** 90bhp (net) at 6500rpm **Maximum torque** 80lb ft (net) at 4000rpm **Transmission** Four- or five-speed manual gearbox, all-synchromesh, 7.87in (200mm) diameter diaphragm spring clutch **Gear ratios** Fifth 0.91, fourth 1.00, third 1.41, second 2.18, first 3.80, reverse 3.65 **Final drive** 4.1:1 **Front suspension** Independent, double wishbones, coil springs, telescopic dampers, anti-roll bar **Rear suspension** Live axle, parallel trailing arms, Panhard rod, coil springs, telescopic dampers, anti-roll bar **Steering** Worm and roller, 15in (380mm) steering wheel, 34ft (10.4m) turning circle, 2.75 turns lock to lock **Brakes** Fiat-Bendix 8.95in (227mm) discs front and rear, Bonaldi vacuum servo **Wheels** Pressed steel disc, four studs, 5in rim **Tyres** Pirelli Cinturato 165 SR 13 **Length** 162.0in (4115mm) **Width** 65.8in (1671mm) **Height** 52.3in (1328mm) **Wheelbase** 95.0in (2413mm) **Kerb weight** 2074lb (941kg) **Top gear speed per 1000rpm** Fifth 17.6mph (28.3kph), fourth 16.1mph (25.9kph) **Top speeds in gears** Fifth 102mph (164kph), fourth 105mph (169kph), third 80mph (129kph), second 52mph (84kph), first 32mph (51kph) **0-60mph** 12.6sec

SPORT COUPÉ 'BC1' (1969–72) As 'AC' except (but note 'BC' also available with 90bhp/1438cc engine): **Bore x stroke** 80.0mm × 80.0mm (3.15in × 3.15in) **Capacity** 1608cc (97.5cu in) **Compression ratio** 9.8:1 **Fuel system** Weber 40 IDF 10/11 carburettor **Maximum power** 110bhp (net) at 6400rpm **Maximum torque** 101lb ft (net) at 3800rpm **Gear ratios** Fifth 0.83, fourth 1.0, third 1.36, second 2.10, first 3.67, reverse 3.53 **Final drive** 4.3:1 **Kerb weight** 2214lb (1004kg) **Top gear speed per 1000rpm** Fifth 17.5mph (28.2kph) **Top speeds in gears** Fifth 109mph (175kph), fourth 93mph (150kph), third 68mph (109kph), second 44mph (71kph), first 26mph (42kph) **0-60mph** 10.7secs

SPORT COUPÉ 'CC1' (1972–75) As 'BC1' except (but note 'CC' also available with 1592cc/108bhp engine): **Bore x stroke** 84mm × 79.2mm (3.31in × 3.12in) **Capacity** 1756cc (107.2cu in) **Compression ratio** 9.8:1 **Fuel system** Weber 34 DMS carburettor **Maximum power** 118bhp (DIN) at 6000rpm **Maximum torque** 111lb ft (DIN) at 4000rpm **Brakes** Fiat-Bendix 9.08in (230mm) discs front and rear, dual circuit, servo **Length** 164.3in **Width** 65.7in **Height** 52.7in **Wheelbase** 95.3in **Weight** 2205lb (1000kg) **Top speeds in gears** Fifth 107mph (172kph), fourth 95mph (153kph), third 70mph (113kph), second 45mph (72kph), first 26mph (42kph) **0-60mph** 10.5sec

ACKNOWLEDGEMENTS

Special colour photography for this book is by James Mann, who shot various Spiders stocked by DTR of Twickenham, Middlesex (0181 891 4043), an Abarth maintained by Middle Barton Garage (01869 340289) and an early 124 Coupé owned by Tony Clements. DTR and Middle Barton also provided extensive information, as did Phil Ward (of *Auto Italia* magazine) and Michael Manning (ex-Fiat). The main sources of period photographs were Fiat Archivio Storico and Pininfarina in Turin, but further illustration has come from Hugh Bishop, David Hodges, Haymarket Publishing, Phil Ward and Martin Holmes. Thanks also to Keith Bluemel for photo research in Italy, and above all to Peter Chippindale for bringing such a sharp eye to text editing.